Table of Content

The Mysterious World of Fundamental Particles

Cosmic Beginnings

Mohit Joshi

Copyright Notice

Discovery of Fundamental Particles

Basic experimental tools for the study of the fundamental particles, their interactions, and composite particles produced through those interactions are particle accelerators, colliders and detectors.

In the particle accelerators, electrons, positrons, protons, antiprotons, etc. are accelerated to the velocity nearly equal to that of light so they acquire extremely high kinetic energy. These high energy electrons are made to collide with high energy positrons coming from the opposite direction or high energy protons are made to collide with high energy protons or antiprotons coming from the opposite direction and the collision may result in the production of massive or high energy particles, i.e. those having rest mass much larger than the rest mass of the electron (0.511 mega electron volt or 0.511 MeV) and the proton (938.272 MeV). Being massive, these particles are highly unstable unlike electrons and protons, and decay rapidly into lighter or stable particles.

These unstable high energy particles are made to pass through the detectors, so that they leave their traces in detectors before decaying or their decay products leave their traces in the detectors. Detectors help find the mass and other properties of these particles.

Collisions of electrons, positrons, protons, antiprotons, and other particles at high energy resulted in the discovery of the fundamental particles and about one hundred resonance particles composed of the fundamental particles.

Classification of Fundamental Particles

The fundamental particles are classified according to their spin. Spin is the intrinsic property of the fundamental particles. It may be thought of as the rotation of a particle on its axis, just as the Earth rotates on its axis, hence the name spin.

On the basis of the spin, the fundamental particles are classified into two groups:
The fermions which are spin-half particles (spin quantum number s = 1/2).
The bosons which are spin-one particles (spin quantum number s = 1).

note: Due to the rotation of a fundamental particle on its axis, it has an intrinsic angular momentum. **A fermion is a spin half particle means the magnitude of this angular momentum** (called the z-component of angular momentum) **along the direction of the motion** (which is along the +z-axis) **is $S_z = +\hbar/2$** (spin 'up', i.e. the direction of the angular momentum is along the direction of the motion of the particle) **or $S_z = -\hbar/2$** (spin 'down', i.e. the direction of the angular momentum is at 180^0 with respect to the direction of the motion of the particle).
There are no spinless fermions.

The fermions are further classified into two groups: The quarks and the leptons. There are six quarks and six leptons.

As per Einstein's equation $E = mc^2$, mass of a particle $m = E/c^2$, and in high energy physics or particle physics, where energy of the particles is frequently measured, mass is defined in terms of energy i.e eV.

1 eV is the energy acquired by an electron, when accelerated through a potential difference of 1 volt.

Thus, mass is defined in terms of 'electron volt eV/c^2', 'mega electron volt MeV/c^2', 'giga electron volt GeV/c^2' or 'tera electron volt TeV/c^2'. (1 MeV = 10^6 eV, 1 GeV = 1000 MeV = 10^9 eV, 1 TeV = 1000 GeV = 10^{12} eV).

Quarks

Six flavours of the quarks (in order of the increasing mass) are: up u, down d, strange s, charm c, bottom b and top t.

Each of u c t quarks has electric charge: +2/3e.

Each of d s b quarks has electric charge: -1/3e.

The rest mass of down quark, up quark, strange quark, charm quark, bottom quark, top quark are 5 MeV, 2 MeV, 100 MeV, 1300 MeV, 4200 MeV, 173.3 GeV respectively.

First generation up quark (charge +2/3e) can decay into a first generation down quark (charge -1/3e) and the down quark can decay into an up quark as the difference between their masses are less.

Second generation charm quark (charge +2/3e) can decay into a second generation strange quark (charge -1/3e) but the decay of the strange quark into charm quark is kinematically forbidden.

Third generation top quark (charge +2/3e) can decay into a third generation bottom quark (charge -1/3e) but the decay of the bottom quark into a top quark is kinematically forbidden.

Whenever during the collisions of electrons, positrons, etc., unstable particles having heavy quarks (s, c, b, t) are produced, they decay into particles having only lighter quarks (u and d). Most of the visible matter in the universe is made up of these lighter and stable up and down quarks, whereas particles having heavier quarks are either produced during the high energy collisions as in particle accelerators or they are found in cosmic rays.

Main properties of the quarks are: mass, spin, parity, electric charge, baryon number and five flavour quantum numbers (isospin, charmness, strangeness, topness and bottomness).

Each of the up quark u, charm quark c and top quark t has +2/3 unit of electric charge, and each of the down quark d, strange quark s and bottom quark b has -1/3 unit of electric charge.

All six quarks have positive or even intrinsic parity, i.e. P = +1.

For all six quarks, baryon number B = +1/3.

The up and down quarks have isospin +1/2 and -1/2 respectively, and all the other four quarks have isospin zero.

The charm quark has charmness quantum number **C = +1,** and all the other five quarks have charmness quantum number C = 0.

The strange quark has strangeness quantum number **S = -1,** and all the other five quarks have strangeness quantum number S = 0.

The top quark has topness quantum number **T = +1,** and all the other five quarks have topness quantum number T = 0.

The bottom quark has bottomness quantum number **B` = -1,** and all the other five quarks have bottomness quantum number B` = 0.

Leptons

Six flavours of the leptons are: electron e^-, muon μ^-, tauon τ^-, electron-type neutrino v_e, muon-type neutrino v_μ and tauon-type neutrino v_τ.

The rest masses of the electron e^-, muon μ^- and tauon τ^- are 0.510998928 MeV, 105.6583715 MeV and 1776.82 MeV respectively, and neutrinos are lighter.

Each of the three charged leptons: electron e^-, muon μ^- and tauon τ^- carries -1 unit of electric charge, whereas corresponding neutrinos are neutral.

Each of the six leptons has the lepton number L = + 1.

Out of three charged leptons, electron (rest mass 0.511 MeV) is a stable and well-known particle.

The tauon (rest mass 1777 MeV) is unstable and decays in $\tau = 2.9 \times 10^{-13}$ seconds.

The muon (rest mass 105 MeV) is also unstable with mean lifetime of 2.2×10^{-6} seconds. On the Earth, the source of the muon is mainly cosmic rays.

note:

The up quark **u** and the down quark **d** are the **first generation quarks.**

The electron **e^-** and the electron-type neutrino **v_e** are the **first generation leptons.**

The charm quark **c** and the strange quark **s** are the **second generation quarks.**

The muon **μ^-** and the muon-type neutrino **v_μ** are the **second generation leptons.**

The top quark **t** and the bottom quark **b** are the **third generation quarks.**

The tauon τ^- and the tauon-type neutrino v_τ are the **third generation leptons.**

Antiparticles

The quarks and the leptons have the corresponding antiparticles too.
Six flavours of the antiquarks (in order of the increasing mass) are: antiup u`, antidown d`, antistrange s`, anticharm c`, antibottom b` and antitop t`.

The quark Q and the corresponding antiquark Q` have the same mass. The magnitude and the direction of the spin are also the same. However, electric charge, colour charge, the weak charge, baryon number, and flavour quantum numbers for the antiquark are of opposite signs. For example, the up and down quarks have electric charge +2/3e and -1/3e respectively and isospin +1/2 and -1/2 respectively, so the antiup quark u` and the antidown quark d` have electric charge -2/3e and +1/3e respectively and isospin -1/2 and +1/2 respectively. The strange quark has strangeness $S = -1$, so the antistrange quark s` has strangeness $S = +1$.

Six flavours of antileptons are: positron e^+, antimuon μ^+, antitauon τ^+, electron-type antineutrino v_e`, muon-type antineutrino $v_\mu$` and tauon-type antineutrino v_τ`.

Each of three charged antileptons: positron e^+, antimuon μ^+ and antitauon τ^+ carries +1 unit of electric charge.
Each of the six antileptons has the lepton number L = - 1.

Bosons

The fundamental bosons are gluon, photon, W^{+-} & Z^0 bosons and graviton.

The gluon and the photon are massless. The W^{+-} and Z^0 bosons have rest mass 80 and 91 GeV respectively.

The gluon, the photon, the W^{+-} and Z^0 bosons are spin one particles, the graviton is spin two particle.

The gluon as well as the photon has negative or odd intrinsic parity, i.e. P = -1.

The graviton has positive or even intrinsic parity, i.e. P = +1.

For W^{+-} and Z^0 bosons which are spin one particles, spin S_z = $+\hbar$, 0 or $-\hbar$.

Since photon is massless, its spin has only two possible values S_z = $+\hbar$ or $-\hbar$.

note: In the standard model of particle physics, there are **12 fundamental fermions** (six quarks and six leptons) and **12 fundamental bosons** (eight gluons, a photon and W^+, W^-, Z^0 bosons) which are the mediators of the interactions between the fermions.

In the standard model, the neutrinos are known to exist in one helicity state and are assumed therefore to be massless. However, the solar neutrino deficit and atmospheric neutrino anomaly are suggestive of neutrino flavour oscillations and therefore of neutrino masses. (See the topic: Neutrinos Oscillation)

Interactions

Each fundamental interaction has an exchanged virtual boson or gauge boson.

These bosons are the mediators of the fundamental interactions which are of four types:

The strong interaction which is mediated by the exchange of a gluon.

The electromagnetic interaction which is mediated by the exchange of a photon.

The weak interaction which is mediated by the exchange of a W^+ (W-plus), or a W^- (W-minus), or a Z^0 (Z-naught or Z-zero) boson.

The gravitational interaction which is mediated by the exchange of a graviton.

The colour charge g_s in a particle causes it to engage in a strong interaction/decay.

The electric charge e in a particle causes it to engage in an electromagnetic interaction/decay.

The weak charge g_w in a particle causes it to engage in a weak interaction/decay.

The **quarks** carry the colour charges, the electric charges and the weak charges that is, all of these three charges, hence they take part in all these interactions/decays.

The **charged leptons:** electron e^-, muon μ^-, tauon τ^- **do not carry colour charges,** so they cannot take part in the strong interactions/decays but they have electric and weak charges so they take part in the electromagnetic and weak interactions/decays.

The neutral leptons or **neutrinos do not carry colour and electric charges,** so they cannot take part in the strong and electromagnetic interactions/decays but they carry weak charges, so they take part in the weak interactions/decays.

Just as electric charges are of two types: positive and negative, colour charges are of six types: red, green, blue, antired, antigreen and antiblue. This has nothing to do with the real colours. The quarks with the same colour charges repel each other and those with different colour charges attract each other. The interquark force is independent of the colours involved.

Each of the six quarks can possess the colour charge in any one of these three forms: red r, green g or blue b. The antiquarks possess anticolour charges: antired r`, antigreen g` and antiblue b`.

The photon which is the mediator of the electromagnetic interaction is uncharged and therefore has no self interaction. The gluon, however, itself carries a net colour charge and therefore has self interaction, i.e. a gluon can couple to the other gluon, i.e. a gluon can emit or absorb a gluon

Whereas the electromagnetic force between electrically charged particles decreases with the increase of the distance between the particles, **the strong force between the quarks, for a very short range, increases as the distance between them increases.**

The strong interaction is about 100 times stronger than the electromagnetic interaction and is about 10^7 times stronger than the weak interaction.

Centre-of-mass Energy

Suppose an incident particle of mass m_1 and total energy E_1 hits a target particle of mass m_2 and energy E_2. If the target particle m_2 is at rest in the laboratory system, then the centre-of-mass energy E_{cms} is approximately equal to $(2m_2E_1)^{1/2}$. If the incident and the target particles travel in the opposite directions, as would be the case in an electron–positron (e^+e^-) or proton-antiproton $(pp^`)$ collider, then E^2_{cms} is approximately equal to $4E_1E_2$, if mass m_1, m_2 are negligible in comparison with E_1, E_2. Thus, the cms energy E_{cms} available for a new particle production in a collider with beams of equal energies E coming from the opposite directions is equal to $2E$ ($E^2_{cms} = 4E^2$) that is, **the cms energy (2E) of the two colliding particles is equal to the sum of their energies (E + E)** whereas for a fixed target machine, the cms energy is proportional to the square root of the incident energy E.

Thus, the highest possible energies for the production of the new particles are found at colliding-beam accelerators. These accelerators accelerate two beams which move in the opposite directions and then are made to collide.

note: Collisions of e^+e^- or $pp^`$ in colliders imply collisions of billions of particles. For example, a beam consisting of 40 bunches (successive bunches being separated by say, 5 meters, the spacing being determined by the radio frequency), each bunch having 10 billion accelerated particles (travelling with the velocity nearly equal to that of light) and each particle having energy of 1000 GeV may be collided with the similar beam coming from the opposite direction in a real experiment in laboratories to discover a massive particle.

Virtual Photons

The electromagnetic interaction takes place only between particles that possess electric charges. Electrically charged particle may be thought of as continuously emitting photons and then reabsorbing them. If another charged particle is nearby, then the photon can be absorbed by it. Photons which we usually see have zero mass. However, the exchanged photons cannot have zero mass. Thus, they are called virtual photons. Exchanged bosons are often called gauge bosons. **Virtual photons are the exchanged bosons or gauge bosons that mediate the electromagnetic interactions.** Virtual photons cannot exist as free particles and therefore are absorbed immediately.

The annihilation of an electron e^- and a positron e^+ to a muon μ^- and an antimuon μ^+ is an example of the electromagnetic interaction. Here, during the collision of the electron e^- and the positron e^+, e^+e^- pair transforms into a virtual photon. The virtual photon then decays into a $\mu^+\mu^-$ pair.

When an electron and a positron collide head-on (i.e. at 180 degree while coming from the opposite directions), a virtual photon may be produced with all the energy of the annihilation going into the virtual photon. That is, this virtual photon will have energy E_γ equal to the cms energy E_{cms} of the colliding e^+e^- pair ($E_\gamma = E_{cms}$). The momentum p of this photon is zero as it is produced in a head-on collision. Thus, from the equation $E^2 = p^2 + m^2$, $E_\gamma^2 = 0 + M_\gamma^2$ (where M_γ is the mass of this photon). Thus, the mass of the photon is equal to its energy ($M_\gamma = E_\gamma = E_{cms}$) that is, this mass is not zero, so it is a virtual photon (the real photon has zero mass). This virtual photon survives for less than say, 10^{-25} seconds and transforms into a charged particle-antiparticle pair.

Just as a real photon can transform into an electron-positron pair, a virtual photon can also transform into an electron-positron pair but as these virtual photons may have extremely high energy too, they can produce muon-antimuon pairs or quark-antiquark pairs (which are very massive) too.

Electromagnetic Interactions

Basically, the electromagnetic interactions involve electrons, positrons and photons: real as well as virtual ones. **The electric charge in a particle causes it to engage in an electromagnetic interaction through a virtual photon.** As electric charges are carried by quarks too, they can also involve in the electromagnetic interactions.

Moller Scattering or electron–electron scattering ($e^- + e^- \rightarrow e^- + e^-$): In this case, when two electrons coming from the opposite directions collide, one of them **emits a virtual photon** and remains itself as an electron. The other electron **absorbs that virtual photon** and remains itself as an electron. After the exchange of a virtual photon between them, these two electrons move in the opposite directions with respect to each other but at an angle with respect to the path of these electrons before the collision.

Bhabha Scattering or electron–positron scattering ($e^- + e^+ \rightarrow e^- + e^+$): In this case, when an electron and a positron coming from the opposite directions collide, they **annihilate to a virtual photon which** subsequently decays into a new electron-positron pair. These new electron and positron move in the opposite directions with respect to each other but at an angle with respect to the path of the original electron-positron pair before the collision.

note: A virtual photon may also decay into **virtual electron-positron pair which** may then again annihilate to a virtual photon, then this virtual photon may decay into a real electron-positron pair. This is equivalent to vacuum polarization.

OR

Like the Moller scattering, when an electron and a positron coming from the opposite directions collide, the electron (or it may be positron too) **emits a virtual photon** and remains itself as an electron. The positron subsequently **absorbs that virtual photon** and remains itself as a positron. After the exchange of a virtual photon between them, these electron and positron move in the opposite directions with respect to each other but at an angle with respect to the path of this electron-positron pair before the collision.

Pair annihilation ($e^- + e^+ \rightarrow \gamma + \gamma$): In this case, when an electron and a positron coming from the opposite directions collide, they **annihilate to a virtual photon which** subsequently decays into a real photon pair. If the cms energy of the colliding electron-positron pair is sufficiently high, then the virtual photon produced through the annihilation of the electron-positron pair would be of high energy and such a virtual photon may decay into a pair of muons: $e^- + e^+ \rightarrow \gamma \rightarrow \mu^- + \mu^+$. If the cms energy of colliding e^+e^- pair is even larger, then the virtual photon would be even more massive and may decay into a pair of tauons or a pair of quarks: $e^- + e^+ \rightarrow \gamma \rightarrow \tau^- + \tau^+$ or $e^- + e^+ \rightarrow \gamma \rightarrow Q + Q`$.

OR

When an electron and a positron coming from the opposite directions collide, **the electron emits a virtual electron** and therefore emits one unit of negative charge, and transforms into a real photon γ (which has zero electric charge). The positron having one unit of positive charge then absorbs this virtual electron and therefore absorbs one unit of negative charge, and transforms into a real photon γ having zero electric charge.

Pair production ($\gamma + \gamma \rightarrow e^- + e^+$): In this case, when the two photons coming from the opposite directions collide, **one** of them **emits a virtual**

electron and therefore emits one unit of negative charge, and transforms into a real positron e^+ having one unit of positive charge. The other photon then **absorbs this virtual electron** and therefore absorbs one unit of negative charge, and transforms into a real electron e^- having one unit of negative charge.

Compton Scattering ($e^- + \gamma \rightarrow e^- + \gamma$): In this case, when an electron and a photon coming from the opposite directions collide, the **electron emits a virtual electron e^-** and therefore emits one unit of negative charge, and transforms into a real photon γ having zero electric charge. The photon which had collided with the electron absorbs the virtual electron and in this way, this photon (which has zero electric charge) absorbs one unit of negative charge and transforms into a real electron e^- having one unit of negative charge.

Also, electron and photon may annihilate to a **virtual electron which** may then decay into an electron and a photon.

note: Compton Scattering is the **inelastic** scattering of a photon by an electron that is, the energy of the photon decreases as part of the energy of the photon is transferred to the electron. The collision causes the electron to recoil and a new photon of lesser energy emerges at an angle from the photon's incoming path).

note: When an electron and a positron collide, they can scatter elastically: $e^- + e^+ \rightarrow e^- + e^+$. The kinetic energy is conserved before and after the elastic collision. The word 'elastic collision' in particle physics means that the same particles came out as went in, i.e. the same particles which had collided, exist after the collision. In the elastic collision, rest energies and masses of the colliding particles are conserved before as well as after the collision.

Gluons and Colour Charges

In the strong interactions, gluons are exchanged between quarks.
The gluons are bicoloured, carrying one unit of positive colour charge and one unit of negative colour charge.

There are eight types of gluons: rb`(red, antiblue), rg`(red, antigreen), bg`, br`, gr`, gb`, $(rr` - bb`)/2^{1/2}$, $(rr` + bb` - 2gg`)/6^{1/2}$.

The gluon may decay into an up quark u and an antiup quark u`.
The gluon may decay into a down quark d and an antidown quark d`.

Similarly, a gluon may decay into a strange quark s and an antistrange quark s`, into a charm quark c and an anticharm quark c`, into a bottom quark b and an antibottom quark b`, into a top quark t and an antitop quark t`.

If **rg`(red antigreen) gluon** (electrically neutral) decays into an up quark u (electric charge: +2/3e) and an antiup quark u`(electric charge: -2/3e), then the up quark will have **red colour charge** and the antiup quark will have **antigreen colour charge,** so that colour charge remained conserved in the decay of a gluon into a quark-antiquark pair. Similarly, if **gb`(green, antiblue) gluon** decays into a strange quark s (charge: -1/3e) and an antistrange quark s` (charge: +1/3e), then the strange quark will have **green colour charge** and the antistrange quark will have **antiblue colour charge.**

Similarly, an up quark u (charge: +2/3e) carrying **blue colour charge** and an antiup quark u` (charge: -2/3e) carrying **antired colour charge** may annihilate to a **br`(blue, antired) gluon** (electric charge: zero), so that **colour charge and electric charge remained conserved** in the annihilation. The **br` (blue, antired) gluon** may then materialize into, say a down quark d (charge: -1/3e) having **blue colour charge** and an antidown quark d` (charge: +1/3e) having **antired colour charge**.

Whereas many particles carry electric charge, no naturally occurring particles carry colour charge.
The only colourless combinations of the quarks are the baryons and the only colourless combinations of the quarks and the antiquarks are the mesons. By colourless, we mean, either the total amount of each colour is zero, as in mesons or all these three colours are present in equal amounts, as in baryons.

Thus, for example, if the up quark u (charge: +2/3e) of the positive pion π^+ (ud`) has **blue** colour charge, then the antidown quark d` (charge: +1/3e) of the positive pion must have **antiblue** colour charge, so that the net colour charge is zero.
Similarly, if the down quark d of a proton p (uud) has **red** colour charge, then one up quark u of the proton must have **blue** colour charge and the other up quark u of the proton must have **green** colour charge.

A red quark (u, d, s, c, b or t), i.e. a quark having red colour charge carries **one unit of redness,** zero blueness, and zero greenness. The corresponding antired quark (u`, d`, s`, c`, b` or t`), i.e. a quark having antired colour carries **minus one unit of redness,** zero blueness, and zero greenness.

Example of decay through a gluon:
The decay of the neutral delta baryon Δ^0 (udd) into a proton p (uud) and a negative pion π^- (u`d):
Δ^0 (udd) $\rightarrow$ p (uud) + π^- (u`d)

Here, one down quark of the neutral delta baryon **emits a gluon** and remains itself as a **down quark d**.

The **gluon then decays into an up quark u and an antiup quark u`.**

This **antiup quark u`** combines with that **down quark d** of the neutral delta baryon which had emitted the gluon and produces a **negative pion $\pi^-$ (u`d).**

The **up quark u** (produced by the decay of the gluon) combines with the up quark and the remaining down quark of the neutral delta baryon to produce a **proton p (uud).**

note: The mass (1232 MeV) of the neutral delta baryon Δ^0 **(udd)** > mass (938 272 MeV) of the proton **p (uud)** + mass (139 MeV) of the negative pion π^- **(u`d).** Thus, this decay is kinematically allowed.

W Boson and Weak Processes

As the leptons and the quarks carry weak charges too, they continually emit and absorb virtual W^{+-} and Z^0 bosons which could be absorbed by a nearby lepton or quark. The exchanged virtual W^{+-} and Z^0 bosons behave like exchanged virtual photons of the electromagnetic interactions but the real W^{+-} and Z^0 bosons are very massive which make the interactions between the quarks and the leptons involving W^{+-} and Z^0 bosons much weaker, hence it is called the weak interaction. Virtual W^{+-} and Z^0 bosons can exist for 3.1×10^{-25} and 2.6×10^{-25} seconds respectively, and the range of the force is only 10^{-17} meters.

In **charged current weak interactions/decays,** the up quark (charge: +2/3e) can transform into a down quark (charge: -1/3e) by emitting a W^+ boson or by absorbing a W^- boson. (Through this emission of the W^+ boson or absorption of the W^- boson, the up quark emits one unit of positive charge or absorbs one unit of negative charge). Similarly, the down quark (charge: -1/3e) can transform into an up quark (charge: +2/3e) by emitting a W^- boson or by absorbing a W^+ boson. (Through this emission of the W^- boson or absorption of the W^+ boson, the down quark emits one unit of negative charge or absorbs one unit of positive charge).

For example, the neutron (udd) can transform into a proton (uud) by emitting a W^- boson through its one down quark d (charge: -1/3e) which will become the up quark u (charge: +2/3e) and then that W^- boson may decay into an electron e^- and electron-type antineutrino $v_e{}^`$.

$W^- \rightarrow e^- + v_e{}^`$

The **strange quark s** (charge: -1/3e) can emit one unit of negative charge by emitting a W⁻ boson (which carries one unit of negative charge) and **transform into an up quark u** (charge +2/3e).

The W⁻ boson can decay into an antiup quark u` (charge: -2/3e) and a down quark d (charge: -1/3e).

W⁻ → u`d

The **antistrange quark s`** (charge: +1/3e) can emit one unit of positive charge by emitting a W⁺ boson (which carries one unit of positive charge) and **transform into an antiup quark u`** (charge -2/3e).

The W⁺ boson can decay into an up quark u (charge: +2/3e) and an antidown quark d`(charge: +1/3e).

W⁺ → ud`

Lepton conservation requires that the total lepton number must be the same on both sides, hence both a neutral and a charged lepton of the same flavour (e.g. electron type antineutrino v_e` with electron e⁻, muon type antineutrino $v_\mu$` with muon μ⁻) to appear together. Thus, when a W⁻ boson (lepton number L = 0) decays into an electron e⁻ (lepton number L = +1), an electron-type antineutrino v_e` (lepton number L= -1) is produced too, so that the lepton number is conserved.

W⁻ → e⁻ + v_e`

When a W⁻ boson (lepton number L = 0) decays into a muon μ⁻ (lepton number L = +1), a muon-type antineutrino v_μ` (lepton number L= -1) is produced too, so that the lepton number is conserved:

W⁻→ μ⁻ + v_μ`

When a W^+ boson (lepton number L = 0) decays into an antimuon μ^+ (lepton number L= -1), a muon-type neutrino v_μ (lepton number L= +1) is produced too.

$$W^+ \rightarrow \mu^+ + v_\mu$$

Similarly,

$$W^+ \rightarrow e^+ + v_e$$
$$W^- \rightarrow \tau^- + v_\tau`$$
$$W^+ \rightarrow \tau^+ + v_\tau$$

W^{+-} and Z^0 bosons (rest mass 80 and 91 GeV) being very massive bosons have a very short lifetime. This means W^{+-} bosons and Z^0 bosons are not directly observed in general. Instead, their decay products are measured.

The W^+ boson decays into a charged antilepton and corresponding neutrino or a quark-antiquark pair, e.g. $W^+ \rightarrow \mu^+ + v_\mu$ or $W^+ \rightarrow ud`$. **The $W^-$ boson decays into a charged lepton and corresponding antineutrino or a quark-antiquark pair,** e.g. $W^- \rightarrow e^- + v_e`$ or $W^- \rightarrow u`d$.

Neutral current weak interactions/decays are mediated by the neutral Z^0 boson.

For example, electron-positron pair e^+e^- annihilates to a Z^0 boson and the Z^0 boson subsequently decays into muon-antimuon pair $\mu^+\mu^-$:

$$e^- + e^+ \rightarrow Z^0 \rightarrow \mu^- + \mu^+.$$

The Z^0 boson (rest mass 91 GeV) being highly unstable immediately decays after its production into fermion-antifermion pair. For example, the Z^0 boson may decay into electron-positron pair e^+e^- or muon-antimuon pair $\mu^+\mu^-$:

$Z^0 \to e^- + e^+$ or $Z^0 \to \mu^- + \mu^+$.

Not only is the the lepton number L conserved in an interaction/decay, the electron number L_e, the muon number L_μ, and the tauon number L_τ are also conserved.

The electron e^- and the electron-type neutrino ν_e are assigned the electron number $L_e = +1$.

The muon μ^- and the muon-type neutrino ν_μ are assigned the muon number $L_\mu = +1$.

The tauon τ^- and the tauon-type neutrino ν_τ are assigned the tauon number $L_\tau = +1$

Corresponding antiparticles are assigned $L_e = -1$ (e^+ and $\nu_e`$), $L_\mu = -1$ ($\mu^+$ and $\nu_\mu`$) and $L_\tau = -1$ (τ^+ and $\nu_\tau`$).

Thus, for example, W^- boson cannot decay into e^- and $\nu_\mu`$ ($W^- \to e^- + \nu_\mu`$ is forbidden) that is, a W^- boson will always emit an electron-type antineutrino $\nu_e`$ with an electron $e^-$ ($W^- \to e^- + \nu_e`$). Thus, the electron number $L_e = 0$ before the decay of W^- boson and $L_e = +1 - 1 = 0$ after the decay too.

Similarly, if an electron e^- emits a W^- boson or absorbs a W^+ boson, it will always convert into an electron-type neutrino ν_e ($e^- \to W^- + \nu_e$ or $e^- + W^+ \to \nu_e$) that is, electron e^- cannot convert into a muon-type neutrino ν_μ ($e^- \to W^- + \nu_\mu$ as well as $e^- + W^+ \to \nu_\mu$ is forbidden) or tauon-type neutrino ν_τ ($e^- \to W^- + \nu_\tau$ as well as $e^- + W^+ \to \nu_\tau$ is forbidden).

Similarly, if a muon μ^- emits a W^- boson or absorbs a W^+ boson, it will always convert into a muon-type neutrino ν_μ ($\mu^- \to W^- + \nu_\mu$ or $\mu^- + W^+ \to \nu_\mu$) and if a tauon τ^- emits a W^- boson or absorbs a W^+ boson, it will always convert into a tauon-type neutrino ν_τ ($\tau^- \to W^- + \nu_\tau$ or $\tau^- + W^+ \to \nu_\tau$).

The **electron-type neutrino v_e neither can emit a W⁻ boson nor absorb a W⁺ boson** to convert into a positron e⁺, i.e. the processes: v_e → W⁻ + e⁺ or v_e + W⁺ → e⁺ are forbidden as the lepton number is not conserved. L = +1 and -1 before and after these processes.

However, **electron-type neutrino v_e can emit a W⁺ boson or absorb a W⁻ boson** to convert into an electron e⁻, i.e. the processes: v_e → W⁺ + e⁻ or v_e + W⁻ → e⁻ are allowed as the lepton number is conserved. L = +1 before and after these processes.

Similarly, **muon-type neutrino $v_μ$ and tauon-type neutrino $v_τ$ can emit a W⁺ boson or absorb a W⁻ boson** to convert into a muon μ⁻ ($v_μ$ → W⁺ + μ⁻ or $v_μ$ + W⁻ → μ⁻) and tauon τ⁻ ($v_τ$ → W⁺ + τ⁻ or $v_τ$ + W⁻ → τ⁻) respectively.

Similarly, an **electron-type antineutrino v_e` or muon-type antineutrino $v_μ$` or tauon-type antineutrino $v_τ$` can emit a W⁻ boson or absorb a W⁺ boson,** to convert into a positron e⁺ ($v_e$` → W⁻ + e⁺ or v_e` + W⁺ → e⁺) or antimuon μ⁺ ($v_μ$` → W⁻ + μ⁺ or $v_μ$` + W⁺ → μ⁺) or antitauon τ⁺ ($v_τ$` → W⁻ + τ⁺ or $v_τ$` + W⁺ → τ⁺) respectively.

The first experimental test of the separate conservation of the electron number and the muon number was conducted at Brookhaven in 1962. About 10^{14} muon-type antineutrinos $v_μ$` were collided with the protons p, then 29 antimuons μ⁺ were detected but no positron e⁺ was detected, i.e. the interaction **$v_μ$` + p → μ⁺ + n** occurred (muon number $L_μ$ = -1 before as well as after this interaction). However, the interaction $v_μ$` + p → e⁺ + n did not occur that is, the interaction: $v_μ$` + p → e⁺ + n is forbidden by muon and electron number conservation as the muon number $L_μ$ = -1 and 0 before and after this interaction and the electron number L_e = 0 and -1 before and after this interaction that is, L_e and $L_μ$ are not conserved in this interaction.

Examples of Charged current weak interactions:

1. Collision between a muon-type antineutrino $v_\mu\grave{}$ and a proton p (uud) may result in the production of an antimuon μ^+ and a neutron n (udd):

$v_\mu\grave{}$ + p (uud) → μ^+ + n (udd)

In this decay, the muon-type antineutrino $v_\mu\grave{}$ with zero electric charge emits one unit of negative electric charge (-1e) by emitting a **W⁻ boson** and transforms into the corresponding charged antilepton, i.e. **antimuon μ^+** with electric charge: +1e.

Then, one of the two up quarks (charge: +2/3e) of the proton (uud) absorbs that W⁻ boson and transforms into a down quark (charge: - 1/3e) and the proton becomes a neutron **n (udd).**

2. Collision between a muon-type neutrino v_μ and a neutron n (udd) may result in the production of a muon μ^- and a proton p (uud):

v_μ + n (udd) → μ^-+ p (uud)

In this decay, the muon-type neutrino v_μ with zero electric charge emits one unit of positive electric charge (+1e) by emitting a **W⁺ boson** and transforms into the corresponding charged lepton, i.e. **muon μ^-** with electric charge: -1e.

Then, one of the two down quarks (charge: -1/3e) of the neutron (udd) absorbs that W⁺ boson and transforms into an up quark (charge: +2/3e) and the neutron becomes a proton **p (uud).**

Through a W boson, quark having up flavour u can transform into a quark having down flavour d (u → d) and similarly, a lepton having electron flavour e⁻ can change into a lepton having electron-type neutrino flavour v_e (e⁻ → v_e). However, photon, gluon, Z^0 boson cannot change quark or lepton flavour. For example, if an up quark emits a photon or a gluon or Z boson, it will remain the up quark that is, no change in the flavour of the quark. Similarly, if an electron emits a photon or a Z^0 boson, it will remain the electron that is, no change in the flavour of the lepton.

Thus, only charged current weak interaction/decay can change the flavour of a fermion (quark or lepton).

The quarks generations are 'skewed' in weak interactions/decays, i.e. 'cross-generational' transitions could occur for the quarks in the weak interactions/decays. In other words, unlike electron number (which is related to the first generation leptons and antileptons: e^-, v_e, e^+, v_e`), upness-plus-downness (which is related to the first generation quarks and antiquarks: u, d, u`,d`) is not conserved in the weak interactions/decays.

Similarly, unlike muon number, strangeness-plus-charmness is not conserved in the weak interactions/decays.

The **probability with which an up quark u** (charge: +2/3e) **can transform into a down quark d** (charge: -1/3e), by emitting a W^+ boson or by absorbing a W^- boson is $|V_{ud}|^2 = 0.974^2 = 0.948$
Similarly, the **probability with which a charm quark c** (charge: +2/3e) **can transform into a bottom quark b** (charge: -1/3e), by emitting a W^+ boson or by absorbing a W^- boson is $|V_{cb}|^2 = 0.041^2 = 0.0016$

The amplitudes with which u, c, t quarks can convert into d, s, b quarks are given below:

$$V_{ud} = 0.974 \quad V_{us} = 0.225 \quad V_{ub} = 0.004$$
$$V_{cd} = 0.225 \quad V_{cs} = 0.986 \quad V_{cb} = 0.041$$
$$V_{td} = 0.008 \quad V_{ts} = 0.040 \quad V_{tb} = 0.999$$

Note that $|V_{ud}|^2 + |V_{us}|^2 + |V_{ub}|^2 = 1$, $|V_{cd}|^2 + |V_{cs}|^2 + |V_{cb}|^2 = 1$, $|V_{td}|^2 + |V_{ts}|^2 + |V_{tb}|^2 = 1$

$|V_{ud}|^2 + |V_{us}|^2 + |V_{ub}|^2 = 1$ implies if an up quark u (charge: +2/3) emits a W⁺ boson or absorbs a W⁻ boson, it **must** have to transform into a down quark d (charge: -1/3e) or a strange quark s (charge: -1/3e) or a bottom quark (charge: -1/3e), as the **sum of the separate probabilities**: $|V_{ud}|^2$ (which is the probability of an up quark to transform into a down quark), $|V_{us}|^2$ (which is the probability of an up quark to transform into a strange quark) and $|V_{ub}|^2$ (which is the probability of an up quark to transform into a bottom quark) **is one**.

$|V_{ud}|^2 = 0.974^2 = 0.948$ is nearly equal to 1. This implies **most of the time**, in the weak interactions/decays, an up quark will transform into a down quark by emitting a W⁺ boson or by absorbing a W⁻ boson. **(u → d)**
The first generation down quark (5 MeV) and the first generation up quark (2 MeV) have similar masses.

$|V_{us}|^2 = 0.225^2 = 0.050$ is although near to zero but still has some value. This implies **sometimes**, in the weak interactions/decays, an up quark will transform into a strange quark by emitting a W⁺ boson or by absorbing a W⁻ boson. **(u → s)**
The second generation strange quark (100 MeV) is massive than the first generation up quark.

$|V_{ub}|^2 = 0.004^2 = 0.000016$ is almost zero. This implies **very rarely**, in the weak interactions/decays, an up quark will transform into a bottom quark by emitting a W⁺ boson or by absorbing a W⁻ boson. **(u → b)**

The third generation bottom quark (4200 MeV) is extremely massive than the first generation up quark.

Similarly, $V_{cd} = 0.225$, $V_{cs} = 0.986$, $V_{cb} = 0.041$ imply **most of the time,** in the weak interactions/decays, a charm quark will transform into a strange quark **(c → s), sometimes** into a down quark **(c → d)**, and **rarely** into a bottom quark **(c → b)** by emitting a W^+ boson or by absorbing a W^- boson.

Similarly, $V_{td} = 0.008$, $V_{ts} = 0.040$, $V_{tb} = 0.999$ imply **most of the time,** in the weak interactions/decays, a top quark will transform into a bottom quark **(t → b), rarely** into a strange quark **(t → s)**, and **very rarely** into a down quark **(t → d)** by emitting a W^+ boson or by absorbing a W^- boson.

Above values show that the transition between the quarks of different generations are suppressed which is called **CKM suppression** (after Cabibbo, Kobayashi, and Maskawa). The transition between the first and the second generation quarks is **less** suppressed: e.g **(u → s),** but the transition between the first and the third generation quarks is **highly** suppressed: e.g. **(u → b).**

There does not seem any cross-generational mixing among the leptons, so electron number, muon number, tauon number should be separately conserved but neutrino oscillations indicate that this conservation is not absolute. The transition between leptons of different generations rarely exists.

Examples of the decays through W bosons:

1. The decay of the neutron n (udd) into a proton p (uud), an electron e⁻ and an electron-type antineutrino v_e` (β-decay):

n (udd) → p (uud) + e⁻ + v_e` (Branching Ratio: ~ 100 %)

This is a weak decay as the antineutrino is involved in this decay which carries only the weak charge.

In this decay, one **down quark d** (charge: -1/3e) of the neutron emits a W⁻ boson (and therefore emits charge: -1e) and **transforms into** an **up quark u** (charge: -1/3e +1e = +2/3e) and the neutron transforms into a **proton p (uud).**

The W⁻ boson subsequently decays into an **electron e⁻** and an **electron-type antineutrino v_e`.**

note:

The neutron n (udd) can decay into a proton p (uud), an electron e⁻ and an electron-type antineutrino v_e` because the mass of the neutron n (939.6 MeV) > mass of the proton p (938.272 MeV) + mass of the electron e⁻ (0.511 MeV) + mass of the electron-type antineutrino $v_e$` (~ 0 MeV))

The neutron n (udd) cannot decay into a proton p (uud) and a negative pion π⁻ (u`d), since the mass of the neutron n (939.6 MeV) < mass of the proton p (938.272 MeV) + mass of the negative pion π⁻ (139.5 MeV). Thus, this decay is kinematically forbidden. That is, the W⁻ boson cannot decay into an antiup quark u` and a down quark d to produce a negative pion π⁻ (u`d).

2. The decay of the tauon τ⁻ into an electron e⁻, an electron-type antineutrino v_e` and a tauon-type neutrino $v_\tau$ or into a muon μ⁻, a muon-type antineutrino $v_\mu$` and a tauon-type neutrino v_τ or into a negative pion π⁻ (u`d) and a tauon-type neutrino v_τ:

$T^- \rightarrow e^- + v_e` + v_\tau$ or $T^- \rightarrow \mu^- + v_\mu` + v_\tau$ or $T^- \rightarrow \pi^-(u`d) + v_\tau$

In this decay, the tauon τ^- with one unit of negative electric charge emits that one unit of negative electric charge by emitting a W^- boson and therefore transforms into the corresponding electrically neutral lepton that is, a **tauon-type neutrino v_τ**.

The W^- boson subsequently decays into an **electron e^-** and the corresponding antineutrino, i.e. **electron-type antineutrino v_e`** or decays into a **muon** and a **muon-type antineutrino $v`_\mu$** or decays into an **antiup quark u`** (charge: -2/3e) and a **down quark d** (charge: -1/3e) which combine to produce a **negative pion $\pi^-$ (u`d).**

note: In this decay, the lepton number is conserved. The tauon τ^- having lepton number L = +1 decays into e^- (L= +1), v_e` (L=-1) and $v_\tau$ (L= +1) or decays into $\mu^-$ (L= +1), $v`_\mu$ (L= -1) and v_τ (L= +1) or decays into π^- (L= 0) and v_τ (L= +1).

note: The tauon is the only lepton that can decay into the hadrons (baryons and mesons) because its mass (1777 MeV) > masses of many hadrons.

Coupling and Coupling Constants

At low energies (say, 1 MeV to 100 MeV), the **coupling constant** for the electromagnetic interaction is $\alpha_e = e^2/4\pi = 1/137$.

The **coupling constant α_e specifies the strength of the electromagnetic interaction** between 'a particle having electric charge' and a photon, i.e. the strength with which a particle having electric charge emits or absorbs a photon.

At low energies (say, 1 MeV to 100 MeV), the strong coupling constant is about 100 times greater than the electromagnetic coupling constant. Thus, at low energies (say, 1 MeV to 100 MeV), the **coupling constant** for the strong interaction $\alpha_s = g_s^2/4\pi = $ **100 ×** $\alpha_e = 100 \times 1/137 \sim 0.7$
Here **g_s** is defined as the strong or colour charge on a quark. Leptons do not carry strong charge.

The **coupling constant α_s specifies the strength of the strong interaction** between 'a particle having strong or colour charge' and a gluon, i.e. the strength with which a particle having strong charge emits or absorbs a gluon, the electrically neutral, massless carrier of the strong force analogous to the photon in electromagnetic interactions.

At low energies (say, 1 MeV to 100 MeV), the weak charge **g_w** on a fermion is calculated to be 0.65.

Thus, at low energies (say, 1 MeV to 100 MeV), the **coupling constant** for the weak interaction is $\alpha_w = g_w^2/4\pi = $ **$0.65^2/4\pi$** $= 1/29.5$
Here **g_w** is defined as the weak charge on a quark or a lepton.

The **coupling constant α_w specifies the strength of the weak interaction** between 'a particle having weak charge' and a W or Z boson, i.e. the strength with which a particle having weak charge emits or absorbs a W or Z boson, the massive carrier of the weak force analogous to the gluon in QCD (quantum chromodynamics) or the photon in QED (quantum electrodynamics).

note: The coupling constant for an interaction/decay is inversely proportional to the square root of the time taken in that interaction/decay provided the mediator of that interaction/decay is massless.

Typically, the decay through the strong process occurs in 10^{-23} s and the decay through the electromagnetic process occurs in 10^{-19} s.

Thus, if α_s and α_e are the strong coupling constant and the electromagnetic coupling constant respectively, then $\alpha_s/\alpha_e = (10^{-19}/10^{-23})^{1/2}$ or $\boldsymbol{\alpha_s/\alpha_e \sim 100.}$

That is, at low energies, the strong coupling constant is about 100 times greater than the electromagnetic coupling constant.

This also implies, at low energy, the strong interaction is about 100 times stronger than the electromagnetic interaction.

Typically, the decay through the weak process occurs in 10^{-10} s and the decay through the electromagnetic process occurs in 10^{-19} s.

This implies, at low energies, the electromagnetic interaction/decay is about $(10^{-10}/10^{-19})^{1/2}$ or $\boldsymbol{\sim 3.16 \times 10^4}$ times stronger than the weak interaction/decay.

However, at low energies, $\alpha_w/\alpha_e = 137/29.5 = 4.64$ i.e. $\boldsymbol{\alpha_w = 4.64\,\alpha_e}$

That is, at low energies, the weak coupling constant is 4.64 times greater than the electromagnetic coupling constant.

This implies, at low energies, the weak interaction is **intrinsically** 4.64 times stronger the electromagnetic interaction.

$\alpha_w/\alpha_e = 4.64$ also implies $g_w^2/e^2 = 4.64$ or $g_w = (4.64)^{1/2}e = 2.15\ e$ that is, the weak charge is greater than the electric charge.

Although, at low energies, **the weak charge and weak coupling constant are greater than the electric charge and the electromagnetic coupling constant** respectively, the weak interactions are actually ~ **3.16 ×10⁴** times weaker than the electromagnetic interactions.

This is so because **in addition to the weak charge, the weak coupling also depends upon the massive propagators, W and Z bosons** which make the weak interactions much weaker than the electromagnetic interactions.

(Whereas the mediators of the strong and the electromagnetic interactions are massless, the mediators (W and Z bosons) of the weak interactions are very massive.)

At low energies, the effect of massive propagators, W and Z bosons, in the weak interaction/decay is such that they decrease about **4.64 × 3.16 ×10⁴** times the strength of the weak interaction/decay, so that the weak interaction/decay (having weak coupling constant 4.64 times greater than the electromagnetic coupling constant at low energies) becomes about **3.16 ×10⁴** times weaker as compared to the electromagnetic interaction/decay, whose mediator, i.e. photon is massless.

The coupling constants are not in fact constant but depend (logarithmically) on the energy scale, at which measurements are being made, hence coupling constants are called **running coupling constants.**

The electromagnetic coupling constant α_e and the value of electric charge e **increases (very slowly)** with the increase of the energy-momentum transfer (q^2).

If we take α_e = 1/137 at 1 MeV, then near the Z boson masses (q^2 = 91 GeV), the effective value of the electromagnetic coupling constant is $\alpha_e(q^2) = \alpha_e$ (91 GeV) = 1/128.

That is, for electron-positron pair colliding at cms energies of 30 GeV, 50 GeV, 60 GeV, 91 GeV, if the value of electromagnetic coupling is α_e (30), α_e (50), α_e (60), α_e (91) respectively, then α_e (30) < α_e (50) < α_e (60) < α_e (91).

The strong or colour coupling constant α_s and the value of strong or colour charge g_s **decreases** with the increase of the energy-momentum transfer (q^2).

For example, in electron-positron collisions at cms energies of 14 GeV, 22 GeV, 35 GeV, 44 GeV, the values of strong coupling constants have been measured as **α_s (14 GeV)** = 0.170, **α_s (22 GeV)** = 0.151, **α_s (35 GeV)** = 0.145, **α_s (44 GeV)** = 0.139 respectively, i.e. α_s (14 GeV) > α_s (22 GeV) > α_s (35 GeV) > α_s (44 GeV)

The weak coupling constant α_w and the value of weak charge g_w **increases** with the increase of the energy-momentum transfer (q^2) but **even more slowly** than the electromagnetic coupling constant α_e.

Hadrons

As the quarks possess colour charges, they engage in the strong interactions through the exchange of the gluons and the attraction between different quarks due to the strong interactions causes the production of the composite particles which are called the hadrons. **Inside hadrons, a quark emits a gluon which is absorbed by the other quark and vice versa.** This continuous emission and absorption of the gluons inside the hadrons gives rise to the strong force. **There are two types of the hadrons: the baryons and the mesons.**
The force between the two quarks (which is responsible in the first instance for binding the quarks together to make the hadrons) is the strong force which is mediated by the exchange of the gluons between the quarks.

The binding of quarks into hadrons increase their effective mass by about 340 MeV.
Whereas the bare masses of u, d, s, c, b, t quarks are 2, 5, 100, 1300, 4200, 173300 MeV respectively, the effective masses are 336, 340, 486, 1550, 4730, 177000 MeV respectively.
Why do the bare quarks have the particular masses they do, no one knows, where all those masses come from.
Isolated quarks cannot be produced because of the quark confinement that is, if the separation between the two quarks increases, their strong interaction causes a quark to radiate a gluon which subsequently decays into a quark-antiquark pair. Original quarks then combine with them to produce the hadrons instead of being isolated.

note:

The gluons themselves carry colour charges and therefore (like the quarks) cannot exist as isolated particles. We can detect gluons only within the hadrons or in colourless combinations with other gluons (glueballs).

The direct gluon-gluon coupling may result in the formation of the glueballs which are bound states of interacting gluons without any quark.

Baryons and Mesons

A baryon QQQ consists of three quarks and a meson QQ` consists of a quark and an antiquark.

Examples of the baryons are protons and neutrons. A proton p (uud) consists of two up quarks and one down quark and a neutron n (udd) consists of one up and two down quarks.

Examples of the mesons are positive pion π^+ (ud`), negative pion $\pi^-$ (u`d), positive kaon K^+ (us`), negative kaon $K^-$ (u`s) and upsilon meson Υ (bb`).

Each antibaryon is composed of three antiquarks.

note: ud` may be read as u anti-d, u`d as anti-u d, us` as u anti-s, u`s as anti-u s, bb` as b anti-b and so on.

As baryons are composed of three valence quarks, i.e. odd number of quarks, they have odd half-integer spin, hence the **baryons are fermions.**

A baryon will have spin 1/2, if two of the three quarks are spinning (rotating) in the same direction, and hence they have spin in the same direction and the third quark is spinning in the opposite direction, and hence it has spin in the opposite direction (1/2 + 1/2 - 1/2 = 1/2).

A baryon will have spin 3/2, if all the three quarks are spinning (rotating) in the same direction, and hence they have spin in the same direction (1/2 + 1/2 + 1/2 = 3/2).

As mesons are composed of quark-antiquark and quark has spin components $\pm 1/2$, it means mesons have spins $\pm 1/2 \pm 1/2$, i.e. 0, -1, +1, i.e. integral spin, hence the **mesons are bosons.** Thus, mesons are not subject to the exclusion principle which means that there is no limit to the number of mesons that can be squeezed into a small space.

note: When the up quark u and the antidown quark d` in a meson are spinning in the same direction, their spins add (1/2 + 1/2 =1) and such a meson is called positive rho meson $\rho^+$ (ud`). When they rotate in the opposite direction, their spins cancel out (1/2 − 1/2 = 0) and such a meson is called positive pion π^+ (ud`). Spins are like magnets in that like repels like, so quarks prefer to spin in opposite directions. Extra energy is needed to align two quarks to spin in the same direction. In lowest energy configuration for a meson, spin of a quark and antiquark are anti-aligned so such a meson is a spin 0 particle, e.g. positive pion π^+. If spins of the quark and the antiquark are parallel to each other, then such a meson will be a spin 1 particle, e.g. positive rho meson ρ^+ and this positive rho meson ρ^+ having spin 1 is more massive than the positive pion π^+ having spin 0. Similarly, positive sigma-star baryon Σ^{*+} (uus) having spin 3/2 is more massive than the positive sigma baryon Σ^+ (uus) having spin 1/2.

note: Each baryon is assigned a baryon number +1 and each antibaryon is assigned a baryon number -1. Each quark and antiquark has baryon number of +1/3 and -1/3 respectively. For mesons, bosons, leptons, baryon number is zero. Baryon number is a conserved quantity that is, in each interaction/decay, baryon number is the same before as well as after the interaction/decay.

For example, in the decay: n (udd) → p (uud) + e^- + v_e`, baryon number before the decay is +1 as neutron has baryon number +1 and baryon number after the decay is also +1 as proton has baryon number +1, whereas $e^-$ and $v_e$` which are leptons, have baryon number of zero.

Similarly, in the decay: Δ^0 (udd) $\rightarrow$ p (uud) + π^- (u`d), baryon number before as well after the decay is +1.

Fine and Hyperfine Structure

Just as the Earth revolves around the Sun and also rotates on its axis, in a hydrogen atom, the electron revolves around the proton and also rotates on its axis.

A charge moving in a circular path, such as an electron orbiting an atomic nucleus effectively constitutes a current loop and a current loop generates magnetic field. Thus, **the electron orbiting the proton generates a magnetic field.**

When there is a current in a loop then it produces a magnetic dipole and acts as a tiny magnet. Electron orbiting the proton has magnetic dipole moment $\mu_e = -(e/2m_e) \times L$ where $L = mvr$ is the orbital angular moment of electron

Fine structure:

The spin orbit interaction between the electron and the proton in a hydrogen atom that is the electromagnetic interaction between the magnetic moment of electron (which is due to its spin) and the magnetic field generated by the electron's orbit around the proton in a hydrogen atom results in different energy levels **for the electron,** each with its own energy ($E = -13.6/n^2$ eV). **These energy levels are called fine structure.**

The spectral emission occurs when an electron transits or jumps from a higher energy state n to a lower energy state n`. The energy of an emitted photon corresponds to the energy difference between the two states.

Due to the transition between different energy levels, there exist various spectral lines. For the hydrogen atom, transition from level n = 2, 3, 4 .. to level n` = 1 results in Lyman series whose all spectral lines are in ultraviolet region, transition from n = 3, 4, 5, 6 .. to n` = 2 results in Balmer series, whose four spectral lines are in visible region of the spectrum with wavelength between 400 nm (1 nanometer = 10^{-9} meters) and 700 nm. Similarly, many other spectral lines exist for the hydrogen atom due to the fine interaction.

Due to the rotation/precession of the electron on its axis, the electron has an intrinsic angular momentum, also known as spin.

The electron (like all other fermions) is spin half particle that is, it has spin quantum number s = 1/2 or intrinsic angular momentum of ħ /2. Due to this intrinsic angular momentum (spin), the electron has an intrinsic spin magnetic dipole moment.

Similarly, the proton possesses intrinsic spin magnetic dipole moment.

Both the intrinsic angular momentum (spin) and magnetic dipole moments are fundamental properties of particles.

A magnetic dipole (eg. bar magnet, current loop, electron, proton) is essentially a small magnet with a north and south pole. Each dipole creates a magnetic field around itself.

Hyperfine structure:

The spin spin interaction between the electron and the proton in a hydrogen atom results in two energy levels. These energy levels are called hyperfine structure.

That is the electromagnetic interaction between the magnetic moment of electron and the magnetic field generated by proton and the electromagnetic interaction between the magnetic moment of proton and the magnetic field generated by electron in a hydrogen atom results in two energy levels. These energy levels are called hyperfine structure.

The interaction energy (E = - µB) is proportional to the magnetic moments of both particles.

Due to the small values of intrinsic spin magnetic dipole moments of electron and proton, the overall scale of spin-spin interaction turns out to be much smaller than the fine structure, hence energy of these two different levels due to the interaction of these magnetic moments is much smaller than the energy of different levels in fine structure. Thus, these energy levels are called hyperfine structure.

Due to the spin-spin interaction, the ground state (E = -13.6 eV) of hydrogen atom, which would normally be a single energy level, is split into two hyperfine levels and their energies are calculated to be $(1/4) \times 5.82 \times 10^{-6}$ eV and $(-3/4) \times 5.82 \times 10^{-6}$ eV respectively so that the difference between these two energy levels $\Delta E = 5.82 \times 10^{-6}$ eV. The two states come from the fact that both the electron and the proton are spin half particles, so there are two possible states: parallel spin and antiparallel spin. The state with parallel spin is slightly higher in energy (less tightly bound). Thus, a hydrogen atom (in ground state) has actually two different energy levels.

The transition of hydrogen atom from energy level $E = (1/4) \times 5.82 \times 10^{-6}$ eV (in which, both electron and proton have parallel spins) to energy level $E = (-3/4) \times 5.82 \times 10^{-6}$ eV (in which, electron spin is antiparallel to proton spin) **causes** the emission of a photon, whose energy is equal to the difference between these two energy levels, i.e. $\Delta E = 5.82 \times 10^{-6}$ eV. The frequency of such a photon will be $\Delta E/h =$ 1420 Mhz and corresponding wavelength 21 cm (centimeter).

The frequency of this spectral line has been measured experimentally and its value obtained is 1420405751.7667 Hz. It is one of the most accurate measurements in physics.

This famous 1420 MHz (21 cm) radio-frequency spin-flip line has been used to map the distribution of atomic hydrogen on cosmic scale. This 21 cm spectral line of the hydrogen atom penetrates a lot of interstellar dust allowing us to see objects with radio telescopes that are otherwise invisible, since their visible light would not penetrate the dust clouds. With radio telescopes tuned in to 21-cm waves (or 1420 MHz), we can observe the velocities and the location of the concentration of atomic hydrogen gas. By measuring the frequency shift due to the Doppler effect, we can find out about the motion of the gas in the galaxy.

Microwave

$v = 3 * 10^8 - 3*10^{11}$ (300 MHz – 300 GHz)

$\lambda = 1 \underline{mm} - 1 \underline{m}$

Radio or Television waves

$v = 3KHz - 300GHz$

$\lambda = 1 mm - 100 km$

The magnetic interaction due to the 'electric charge and spin' of the electron and the proton in a hydrogen atom results in hyperfine structure having two energy levels, similarly the magnetic interaction due to the 'colour charge and spin' of the quarks inside a baryon or meson, there exists different energy levels.

note:

The magnitude of the intrinsic spin magnetic dipole moment of electron $(\mu_s) = -(e/2m_e) * g_s * S$, where g_s is the g-factor for electron spin (approximately 2), and S is the spin quantum number ($\hbar/2$ for an electron).

Thus, the magnitude of the intrinsic spin magnetic dipole moment of electron $(\mu_s) = -(e\hbar/2m_e) = -\mu_B = -5788 \times 10^{-14}$ MeV/Tesla (Tesla is the unit of magnetic field strength)

This magnetic moment of electron is due to the 'electric charge and spin' of the electron.

The Bohr magneton μ_B is a fundamental constant that serves as a natural unit for expressing magnetic moments of electrons and other particles.

note:

The magnitude of the intrinsic spin magnetic dipole moment of proton $(\mu_p) = +2.79285 \times (e\hbar/2m_p) = +2.79285 \times \mu_N = +2.79285 \times 3.152 \times 10^{-14}$ MeV/Tesla $= 8.79 \times 10^{-14}$ MeV/Tesla which is 658.47 times smaller than the intrinsic spin magnetic dipole moment of the electron.

This magnetic moment of proton is due to the 'electric charge and spin' of the proton.

The nuclear magneton μ_N is 1836 times smaller than the Bohr magneton because the proton is 1836 times heavier than the electron.

The nuclear magneton is used to express the magnetic moments of nucleons (protons and neutrons) and atomic nuclei.

note:

Hyperfine structure can be observed during the **Zeeman effect** which is the **splitting of energy levels** in the presence of external magnetic fields. The splitting of energy levels in the Zeeman effect is directly proportional to the applied magnetic field.

Hyperfine splitting is the process of **splitting spectral lines** into closely spaced lines when an atom is subjected to a varying magnetic field.

Discovery of Partons inside Nucleons

Deep inelastic lepton-nucleon scattering, e.g. electron-proton scattering ($e^- + p \rightarrow e^- + p$) & neutrino-nucleon scattering ($v_\mu + N \rightarrow v_\mu + N$) and **electron-positron annihilation to hadrons ($e^- + e^+ \rightarrow$ hadrons)** provided the evidence that nucleons (protons and neutrons) contain fractionally charged quarks as real dynamical entities.

When **electrons or neutrinos were incident on protons target,** most of the incident particles passed right through, whereas a small number bounced back sharply. This means that the charge of the proton is concentrated in small lumps. The evidence suggested that there are three lumps inside the proton. Those lumps (which we now say quarks) were called the partons.

These experiments also provided the evidence for the existence of neutral gluons among the parton constituents. **Neutral gluon appears in three-jet events in electron-positron annihilation. See the topic: Discovery of Vector Mesons.**

The Pauli Exclusion Principle states that no two identical spin 1/2 particles can occupy the same state. However, double positive delta baryon Δ^{++} (uuu) consists of 3 identical up quarks with parallel spins and all of these three quarks are in the spherically symmetrical ground state (orbital angular momentum $l = 0$). Similarly, omega baryon Ω^- (sss) consists of 3 identical strange quarks with parallel spins. That is, all the three quarks in Δ^{++} (uuu) and Ω^- (sss) have same quantum state which is contrary to the Pauli Exclusion Principle. This implied each of the quarks comes in three colours (red, blue, green) and in a baryon, there exists one quark of each colour. Then, three up quarks in Δ^{++} (uuu) and three strange quarks in Ω^- (sss) are not identical. Thus, Pauli's

Exclusion principle is not violated. **In this way, three different colour charges were discovered.**

Isospin and Strangeness

There are some particles which are composed of the same number of the up and down quarks. The up and down quarks have similar masses and similar strong interactions but the up quark has charge: +2/3e and the down quark has charge: -1/3e.

Thus, the particles composed of the same number of the up and down quarks also have similar masses and similar strong interactions but different electric charges.

The family of hadrons **having** similar masses but different electric charges is called isospin multiplet and assigned isospin quantum number I and each individual charged state has a different isospin projection I_3.

All hadrons within an isospin multiplet have the same spin, parity, and baryon numbers, but differ in electric charges.

Strange quark has strangeness quantum number S = -1 and antistrange quark has strangeness quantum number S = +1. Thus,

If a baryon has zero, one, two, or three strange quarks, it means its strangeness is S = 0, -1, -2, -3 respectively.

If a meson has no strange quark and no antistrange quark, it means its strangeness is S = 0.

If a meson has one strange quark and zero antistrange quark, it means its strangeness is S = -1.

If a meson has one strange quark and one antistrange quark, it means its strangeness is S = 0.

If a meson has one antistrange quark and zero strange quark, it means its strangeness is S = +1.

Based on isospin and strangeness, hadrons (baryons and mesons) are organised into multiplets. These multiplets are called baryon octet, baryon decuplet, pseudoscalar meson nonet and vector meson nonet.

Baryons with spin $s = 1/2$ form the baryon octet, i.e. there are 8 baryons in this multiplet.

Baryons with spin $s = 3/2$ form the baryon decuplet, i.e. there are 10 baryons in this multiplet.

Mesons with spin $s = 0$ form the pseudoscalar meson nonet, i.e. there are nine mesons in this multiplet.

Mesons with spin $s = 1$ form the vector meson nonet, i.e. there are also nine mesons in this multiplet.

The Baryon Octet

The Baryon Octet

(Eight particles each with spin 1/2 and even parity)

n (udd), p (uud)

S = 0, 939 MeV

Isospin doublet

Σ⁻ (dds), Σ⁰ (uds), Λ⁰ (uds), Σ⁺ (uus)

S = -1, 1193 MeV / 1116 MeV

Isospin triplet (Each particle has 1s)

Ξ⁻ (dss), Ξ⁰ (uss)

S = -2, 1318 MeV

Isospin doublet (Each particle has 2s)

In Baryon Octet, there are proton and neutron (each with rest mass of about 939 MeV), 3 Sigma baryons (each with rest mass of about 1193 MeV), 1 Lambda baryon (1116 MeV) and 2 Xi baryons (each with rest mass of about 1318 MeV).

Nucleons, i.e. proton (uud) and neutron (udd) **have** similar masses, similar strong interactions as each of them contains total three up and/or down quarks.

p (uud) and n (udd) having similar masses (939 MeV) but different electric charges form Isospin doublet with isospin $I = 1/2$ and individual charged states, i.e. p (uud) and n (udd) with electric charges +1 and 0 respectively (+2/3 + 2/3 -1/3 = +1 and +2/3 -1/3 -1/3 = 0) **have** Isospin projection $I_3 = +1/2$ and -1/2 respectively.

Proton and neutron contain no strange quark, so they have strangeness S = 0.

3 Sigma baryons Σ^+ (uus), Σ^0 (uds), Σ^- (dds) **have** similar masses, similar strong interactions, as each of them contains total two up and/or down quarks and one strange quark.

Σ^+ (uus), Σ^0 (uds), Σ^- (dds) having similar masses (1193 MeV) but different electric charges form Isospin triplet with Isospin I =1 and individual charge states, i.e. Σ^+ (uus), Σ^0 (uds), Σ^- (dds) with electric charges +1, 0, -1 respectively **have** Isospin projections I_3 = +1, 0, -1 respectively.

Each of Σ^+ (uus), Σ^0 (uds), Σ^- (dds) contains one strange quark, so they have strangeness S = -1.

Lambda baryon – Λ^0 (uds) has electric charge 0, Isospin 0 and strangeness S = -1.

2 Xi baryons Ξ^0 (uss), Ξ^- (dss) **have** similar masses, similar strong interactions, as each of them contains one up and/or down quark and two strange quarks.

Ξ^0 (uss), Ξ^- (dss) having similar masses (1318 MeV) but different electric charges form Isospin doublet with Isospin I =1/2 and individual charge states, i.e. Ξ^0 (uss), Ξ^- (dss) with electric charges 0, -1 respectively **have** Isospin projections I_3 = +1/2 and -1/2 respectively.

Each of Ξ^0 (uss), Ξ^- (dss) contains two strange quark, so they have strangeness S = -2

note: Read Σ^+ as positive sigma baryon or sigma-plus, Σ^0 as neutral sigma baryon or sigma-naught or sigma-zero, Σ^- as negative sigma baryon or sigma-minus.

note: Σ^0 (uds) and Λ^0 (uds) each contains one up, one down and one strange quark but Σ^0 (1192.6 MeV) is more massive than Λ^0 (1115.6 MeV). Σ^0 (1192.6 MeV) is the electromagnetic excited state of Λ^0 (1115.6 MeV). Σ^0 (1192.6 MeV) decays into Λ^0 (1115.6 MeV) and a photon.

note: As an up quark has electric charge: +2/3e and a strange quark has electric charge: -1/3e, thus the positive sigma baryon Σ^+ (uus) consisting of two up quarks and one strange quark has electric charge +2/3e + 2/3e - 1/3e = +1e and the neutral X-baryon Ξ^0 (uss) consisting of one up quark and two strange quarks has electric charge +2/3e - 1/3e - 1/3e = 0. The same thing applies to all other composite particles.

The Baryon Decuplet

The Baryon Decuplet

(Ten particles each with spin 3/2 and even parity)

Δ^- (ddd), Δ^0 (udd), Δ^+ (uud), Δ^{++} (uuu)

S = 0, 1232 MeV

Isospin quartet

Σ^{*-} (dds), Σ^{*0} (uds), Σ^{*+} (uus)

S = -1, 1385 MeV

Isospin triplet

Ξ^{*-} (dss), Ξ^{*0} (uss)

S = -2, 1530 MeV

Isospin doublet

Ω^- (sss)

S = -3, 1672 MeV

In Baryon Decuplet, there are 4 Delta baryons (each with rest mass of about 1232 MeV), 3 Sigma-star baryons (each with rest mass of about 1385 MeV), 2 Xi-star baryons (each with rest mass of about 1530 MeV) and 1 Omega baryon (1672 MeV).

4 Delta baryons Δ^{++} (uuu), Δ^+ (uud), Δ^0 (udd), Δ^- (ddd) have similar masses, similar strong interactions as each of them contains total three up and/or down quarks.

Δ^{++} (uuu), Δ^+ (uud), Δ^0 (udd), Δ^- (ddd) having similar masses (1232 MeV) but different electric charges form Isospin quartet with Isospin I = 3/2 and individual charge states, i.e. Δ^{++} (uuu), Δ^{+} (uud), Δ^{0} (udd), Δ^{-} (ddd) with electric charges +2, +1, 0, -1 respectively have Isospin projections I_3 = +3/2, +1/2, -1/2, -3/2 respectively.

Δ^{++} (uuu), Δ^{+} (uud), Δ^{0} (udd), Δ^{-} (ddd) contain no strange quark, so they have strangeness S = 0.

3 Sigma-star baryons Σ^{*+} (uus), Σ^{*0} (uds), Σ^{*-} (dds) have similar masses, similar strong interactions, as each of them contains total two up and/or down quarks and one strange quark.

Σ^{*+} (uus), Σ^{*0} (uds), Σ^{*-} (dds) having similar masses (1385 MeV) but different electric charges form Isospin triplet with Isospin I =1 and individual charge states, i.e. Σ^{*+} (uus), Σ^{*0} (uds), Σ^{*-} (dds) with electric charges +1, 0, -1 respectively have Isospin projections I_3 = +1, 0, -1 respectively.

Each of Σ^{*+} (uus), Σ^{*0} (uds), Σ^{*-} (dds) contains one strange quark, so they have strangeness S = -1.

2 Xi-star (Ξ^*) baryons Ξ^{*0} (uss), Ξ^{*-} (dss) have similar masses, similar strong interactions, as each of them contains one up and/or down quark and one strange quark.

Ξ^{*0} (uss), Ξ^{*-} (dss) having similar masses (1530 MeV) but different electric charges form Isospin doublet with Isospin I =1/2 and individual charge states, i.e. Ξ^{*0} (uss), Ξ^{*-} (dss) with electric charges 0, -1 respectively have Isospin projections I_3 = +1/2 and -1/2 respectively. Each of Ξ^{*0} (uss), Ξ^{*-} (dss) contains two strange quark, so they have strangeness S = -2

Omega (Ω) particle or Omega baryon – Ω^- (sss) has electric charge -1 and Isospin 0 and strangeness S = -3.

note: A **hyperon** is a baryon containing one, two or three strange quarks, but no charm, bottom, or top quark. Thus, Sigma Baryons, Lambda Baryon, Xi Baryons, Sigma-Star Baryons, Xi-star Baryons and Omega Baryon are Hyperons.

note: Read Σ^{*+} as positive sigma-star baryon or sigma-star-plus, Σ^{*0} as neutral sigma-star baryon or sigma-star-naught or sigma-star-zero, Σ^{*-} as negative sigma-star baryon or sigma-star-minus.
Read Ξ^{*0} as neutral Xi-star baryon or Xi-star-naught or Xi-star-zero, Ξ^{*-} as negative Xi-star baryon or Xi-star-minus.

note: The lambda baryon, each of sigma baryons and each of sigma-star baryons contain one strange quark. Each of Xi baryons and each of Xi-star baryons contain two strange quarks.

note: A particle has its mass as a part of its name, if and only if it decays strongly or electromagnetically. Thus, for example, delta baryons, Σ-star baryons, rho mesons which decay strongly may be written as $\Delta(1232)^{++}$, $\Delta(1232)^{0}$, $\Sigma(1385)^{+}$, $\Sigma(1385)^{0}$, $\Sigma(1385)^{-}$, $\rho(770)^{+}$, $\rho(770)^{0}$ and so on. However, Σ^{-}, Ω^{-}, π^{+}, π^{0}, π^{-}, etc. which decay weakly are simply written as Σ^{-}, Ω^{-}, π^{+}, π^{0}, π^{-}. The neutral charmed omega-star baryon Ω_{c}^{*0} (ssc) decays electromagnetically and is written as $\Omega_{c}(2770)^{0}$.

note:
Δ baryons (1232 MeV) in the baryon decuplet are **more massive** than nucleons (939 MeV) in the baryon octet.
Σ-star baryons (1385 MeV) in the baryon decuplet are **more massive** than Σ baryons (1193 MeV) in the baryon octet.

Ξ-star baryons (1530 MeV) in the baryon decuplet are **more massive** than Ξ baryons (1318 MeV) in the baryon octet.

This is due to a quantum mechanical phenomenon called hyperfine structure.

The spin spin interactions between quarks and/or antiquarks in hadrons results in different energy levels called hyperfine structure.

Nucleons are spin half particles that is, **in each of 2 Nucleons**: p (uud), n (udd), two of the three quarks have parallel spin and the spin of the third quark is antiparallel to the spins of those two quarks and the colour magnetic interaction between these three spins **in each of 2 Nucleons** gives rise to the **low energy level: 939 MeV.**

Delta Baryons are spin 3/2 particles that is, **in each of 4 Delta Baryons**: Δ^{++} (uuu), Δ^{+} (uud), Δ^{0} (udd), Δ^{-} (ddd), all the three quarks have parallel spin and the colour magnetic interaction between these three spins **in each of 4 Delta Baryons** gives rise to the **high energy level: 1232 MeV.**

That is, the colour magnetic interaction **between** the spins of three up and/or down quarks gives rise to two energy levels: 939 MeV (corresponding to nucleons having spin 1/2) and 1232 MeV (corresponding to delta baryons having spin 3/2).

note: The up quark and the down quark have slightly different masses and each of four delta baryons (which are four different particles) have 3, 2, 1, 0 up quarks and 0, 1, 2, 3 down quarks respectively. Thus, all these 4 delta baryons do not have the same rest energy of 1232 MeV. Instead, their masses differ from another by a fraction of a MeV. Similarly, the proton (having 2 up and 1 down quarks) has exact mass of 938.272 MeV and the neutron (1 up and 2 down quarks) has exact mass of 939.565 MeV.

Sigma Baryons are spin half particles that is, **in each of 3 Sigma Baryons:** Σ^+ (uus), Σ^0 (uds), Σ^- (dds), two of the three quarks have parallel spin and the spin of the third quark is antiparallel to the spins of those two quarks and the colour magnetic interaction between these three spins **in each of 3 Sigma Baryons** gives rise to the **low energy level: 1193 MeV.**

Sigma-star Baryons are spin 3/2 particles that is, **in each of 3 Sigma-star Baryons:** Σ^{*+} (uus), Σ^{*0} (uds), Σ^{*-} (dds), all the three quarks have parallel spin and the colour magnetic interaction between these three spins **in each of 3 Sigma-star Baryons** gives rise to the **high energy level: 1385 MeV.**

That is, the colour magnetic interaction **between** the spins of 'two up and/or down quarks and one strange quark' gives rise to two energy levels: 1193 MeV (corresponding to sigma baryons having spin 1/2) and 1385 MeV (corresponding to sigma-star baryons having spin 3/2).

The Pseudoscalar Meson Nonet

The Pseudoscalar Meson Nonet
(Nine particles each with spin 0 and odd parity)

neutral kaon **K⁰ (ds`)**, positive kaon **K⁺ (us`)**
S = +1, 493 MeV
Isospin doublet

π⁻ (u`d), ⁰ {(uu`- dd`)/2^{1/2}}, η {(uu` + dd` - 2ss`)/6^{1/2}}, η' {(uu` + dd` + ss`)/3^{1/2}}, π⁺ (ud`)
S = 0, 134 MeV / 547 MeV / 957 MeV
Isospin triplet (π)

negative kaon **K⁻ (u`s)**, antineutral kaon **K⁰` or anti-K⁰ (d`s)**
S = -1, 493 MeV
Isospin doublet

In Pseudoscalar Meson Nonet, there are 4 Kaons (each with rest mass of about 493 MeV), 3 Pions (each with rest mass of about 134 MeV) and 2 Eta mesons (each with rest mass of about 547 and 957 MeV).

Neutral kaon K⁰ (ds`) contains one antistrange quark. Antineutral kaon K⁰` (d`s) contains one strange quark.

4 Kaons form two doublets of isospin.
2 Kaons, i.e. positive kaon K⁺ (us`), neutral kaon K⁰ (ds`) having similar masses (493 MeV) but different electric charges form Isospin doublet with Isospin I = 1/2 and individual charge states, i.e. K⁺ (us`), K⁰ (ds`)

with electric charges +1, 0 respectively have Isospin projections $I_3 =$ +1/2 and -1/2 respectively and strangeness S = +1.

2 Kaons, i.e. antineutral kaon $K^{0`}$ or anti-$K^0$ (d`s), negative kaon K^- (u`s) having similar masses (493 MeV) but different electric charges form Isospin doublet with Isospin I = 1/2 and individual charge states, i.e. $K^{0`}$ or anti-K^0 (d`s), $K^-$ (u`s) with electric charges 0, -1 respectively have Isospin projections $I_3 =$ +1/2 and -1/2 respectively and strangeness S = -1.

3 Pions: positive pion π^+ (ud`), neutral pion $\pi^0$ {(uu`- dd`)/$2^{1/2}$}, negative pion $\pi^-$ (u`d) having similar masses (134 MeV) but different electric charges form Isospin triplet with Isospin I =1 and individual charge state, i.e. π^+ (ud`), $\pi^0$ {(uu`- dd`)/$2^{1/2}$}, $\pi^-$ (u`d) with electric charges +1, 0, -1 respectively have Isospin projections I_3 = +1, 0, -1 respectively and strangeness S = 0.

Eta meson – η {(uu` + dd` - 2ss`)/$6^{1/2}$} and **Eta prime meson** – η' {(uu` + dd` + ss`)/$3^{1/2}$} have Isospin 0 and strangeness S = 0.

These **meson nonet** consisting of up, down and strange quarks only are **pseudoscalar mesons.** (See the topic: Parity).

note:

Read K^+ (us`) as positive kaon or positive K-meson or K-plus.

K^0 (ds`) as neutral kaon or neutral K-meson or K-naught or K-zero.

$K^{0`}$ (d`s) as antineutral kaon or antineutral K-meson or anti-K-naught or anti-K-zero.

K^- (u`s) as negative kaon or negative K-meson or K-minus.

Read π^+ (ud`) as positive pion or positive pi meson or pi-plus.

π^0 (uu`) as neutral pion or neutral pi meson or pi-naught or pi-zero.

π^- (u`d) as negative pion or negative pi meson or pi-minus.

note:

Each of the positive kaon K⁺ (us`) and the neutral kaon K⁰ (ds`) contains one antistrange quark s`.

Each of the negative kaon K⁻ (u`s) and the antineutral kaon K⁰` (d`s) contains one strange quark s.

The positive kaon K⁺ (us`) and the negative kaon K⁻ (u`s) contains one up quark u and one antiup quark u` respectively.

The neutral kaon K⁰ (ds`) and the antineutral kaon K⁰` (d`s) contains one down quark d and one antidown quark d` respectively.

note: As an up quark has electric charge: +2/3e and a down quark has electric charge: -1/3e, thus the positive pion π⁺ (ud`) consisting of one up quark u (charge: +2/3e) and one antidown quark d` (charge: +1/3e) has electric charge +2/3e + 1/3e = +1e. The same thing applies to all other composite particles.

The Vector Meson Nonet

The Vector Meson Nonet

(Nine particles each with spin 1 and odd parity)

K-star-naught **K*⁰ (ds`)**,  K-star-plus **K\*⁺ (us`)**

S = +1, 892 MeV

Isospin doublet

ρ⁻ (u`d), ρ⁰ {(uu`- dd`)/2¹ᐟ²}, ω {(uu` + dd`)/2¹ᐟ²}, φ (ss`), ρ⁺ (ud`)

S = 0, 770 MeV / 782 MeV / 1020 MeV

Isospin triplet (ρ)

K-star-minus **K*⁻ (u`s)**,  anti-K-star-naught **K\*⁰` or anti-K*⁰ (d`s)**

S = -1, 892 MeV

Isospin doublet

In Vector Meson Nonet, there are 4 K-star mesons (each with rest mass of about 892 MeV), 3 Rho meson (770 MeV), Omega meson (782 MeV) and Phi meson (1020 MeV).

4 K-star mesons form two doublets of isospin.
2 K-star mesons, i.e. K-star-plus K*⁺ (us`), K-star-naught K\*⁰ (ds`) having similar masses (892 MeV) but different electric charges states form Isospin doublet with Isospin I = 1/2 and individual charge states, i.e. K*⁺ (us`), K\*⁰ (ds`) with electric charges +1, 0 respectively have Isospin projections I_3 = +1/2 and -1/2 respectively and strangeness S = +1.

2 K-star mesons, i.e. anti-K-star-naught K*⁰` or anti-K*⁰ (d`s), K-star-minus K*⁻ (u`s) having similar masses (892 MeV) but different electric charges form Isospin doublet with Isospin I = 1/2 and individual charge states, i.e. $K^{*0`}$ or anti-K^{*0} (d`s), $K^{*-}$ (u`s) with electric charges 0, -1 respectively have Isospin projections I_3 = +1/2 and -1/2 respectively and strangeness S = -1.

3 Rho mesons rho plus ρ^+ (ud), rho naught ρ^0 {(uu`- dd`)/$2^{1/2}$}, rho minus ρ^- (u`d) having similar masses (770 MeV) but different electric charges form Isospin triplet with Isospin I =1 and individual charge states, i.e. $\rho^+$ (ud`), ρ^0 {(uu`- dd`)/$2^{1/2}$}, ρ^- (u`d) with electric charges +1, 0, -1 respectively have Isospin projections I_3 = +1, 0, -1 respectively and strangeness S = 0.

Omega meson – ω {(uu` + dd`)/$2^{1/2}$} has Isospin 0 and strangeness S = 0.

Phi meson – φ (ss`) has Isospin 0 and strangeness S = 0.

These **meson nonet with spin 1** consisting of up, down and strange quarks only are **vector mesons.** (See the topic: Parity).

note: Read K^{*+} as positive K-star meson or K-star-plus, K^{*0} as neutral K-star meson or K-star-naught or K-star-zero, $K^{*0`}$ (anti-K^{*0}) as antineutral K-star meson or anti-K-star-naught or anti-K-star-zero, K^{*-} as negative K-star meson or K-star-minus.

note: K-star mesons (892 MeV) in vector meson nonet are **more massive** than kaons (493 MeV) in pseudoscalar meson nonet. Rho mesons (770 MeV) in vector meson nonet are **more massive** than pions (134 MeV) in pseudoscalar meson nonet.

Kaons are spin 0 particles that is, **in each of 4 Kaons:** K^+ (us`), $K^0$ (ds`), $K^{0`}$ (d`s), K^- (u`s), the spin of the quark and the antiquark are antiparallel to each other and the colour magnetic interaction between these two spins **in each of 4 Kaons** gives rise to the **low energy level: 493 MeV.**

K-star mesons are spin 1 particles that is, **in each of 4 K-star meson:** K*+ (us`), K*0 (ds`), K*0` (d`s), K*− (u`s), the spin of the quark and the antiquark are parallel to each other and the colour magnetic interaction between these two spins **in each of 4 K-star mesons** gives rise to the **high energy level: 892 MeV.**

That is, the colour magnetic interaction **between** the spins of up and antistrange quark (us`), down and antistrange quark (ds`), antidown and strange quark (d`s), antiup and strange quark (u`s) gives rise to two energy levels: 493 MeV (corresponding to kaons having spin 0) and 892 MeV (corresponding to K-star mesons having spin 1).

(Actual rest masses of K+, K−, K0, K0` are 493.677 MeV, 493.677 MeV, 497.614 MeV, 497.614 MeV respectively. Actual rest masses of K*+, K*−, K*0, K*0` are 891.66 MeV, 891.66 MeV, 895.81 MeV, 895.81 MeV respectively).

Pions are spin 0 particles that is, **in each of 3 Pions**: π+ (ud`), π0 {(uu`- dd`)/2$^{1/2}$}, π− (u`d), the spin of the quark and the antiquark are antiparallel to each other and the magnetic interaction between these two spins **in each of 3 Pions** gives rise to the **low energy level: 134 MeV.**

Rho mesons are spin 1 particles that is, **in each of 3 Rho meson:** ρ+ (ud`), ρ0 {(uu`- dd`)/2$^{1/2}$}, ρ− (u`d), the spin of the quark and the antiquark are parallel to each other and the magnetic interaction between these two spins **in each of 3 Rho mesons** gives rise to the **high energy level: 770 MeV.**

That is, the colour magnetic interaction **between** the spins of 1 up/down quark and 1 antiup/antidown quark gives rise to two energy levels: 134

MeV (corresponding to pions having spin 0) and 770 MeV
(corresponding to rho mesons having spin 1).

(The up quark and the down quark have slightly different masses. Thus, all these 3 pions do not have the same rest energy of 134 MeV. Actual rest masses of π^+, π^-, π^0 are 139.57018 MeV, 139.57018 MeV, 134.9766 MeV respectively).

note: There is much larger mass splitting (636 MeV) between π (134 MeV) and ρ (770 MeV) as compared with that (293 MeV) between Δ (1232 MeV) and nucleon N (939 MeV). It is known that rms radius of the charge distribution of mesons is smaller (R_0 = 0.6 fm) than it is for baryons (R_0 = 0.8 fm), so that interaction energy which is proportional to $1/R_0^3$, will be $(0.8/0.6)^3$ or ~ 2.3 times larger for meson, hence larger difference in mass splitting between meson and its excited state.

The various energy levels in the hydrogen atom which is an electron/proton system, are relatively close together. The **energy spacing** is typically of some electron volts (fine structure) and micro electron volts (hyperfine structure), so that we think all those energy levels as 'hydrogen'. However, in baryon octet, baryon decuplet, and meson nonets, the **energy spacings** for the **different** different states of a bound quark system are very large (hyperfine structure), hence we regard them as distinct particles.

In baryon octet, baryon decuplet, meson nonets, the quarks do not revolve around one another, thus their orbital angular momentum is zero, so no fine structure exists for them.

The rhos may be thought of as 'excited' pions, the K-star mesons may be thought of as 'excited' koons, the delta baryons may be thought of as 'excited' nucleons and similarly for other particles.

note: $\pi^0 \{(u\bar{u} - d\bar{d})/2^{1/2}\}$ implies if in some decay processes, say, 400 neutral pions are produced, then 200 would be consisting of up and antiup quarks, i.e. $\pi^0 (u\bar{u})$ and remaining 200 would be consisting of down and antidown quarks $\pi^0 (d\bar{d})$. The same thing applies to the neutral rho meson, omega meson, etc.

Cosmic Rays and Muons

Cosmic Rays refer to the elementary particles, nuclei, and electromagnetic radiation of extra-terrestrial origin. These may include short-lived particles such as pions, muons, lambda baryons, etc.

Energy of particles in cosmic rays vary from ~100 MeV up to ~10^{12} GeV. The highest energy of cosmic rays is higher than the best particle accelerator LHC (Large Hadron Collider) can generate.

Composition of cosmic rays changes with energy. For example, at 1 GeV, cosmic rays are composed of ~ 85% proton, ~ 12 % helium, ~ 1% heavy nuclei and ~ 2% electron. In cosmic rays, there exists very small amount of antimatter such as positron e^+ and antiproton p`.

There exist several different magnetic fields from the origin of cosmic rays to Earth: magnetic field of Earth, solar system and galaxy. So unlike neutral particles such as gamma rays, cosmic rays cannot point to the source because cosmic rays are highly charged particles and cannot reach to Earth without losing their original direction as they will be deflected by various magnetic fields.

Primary cosmic rays **originating** from stars, supernova explosions, etc. and arriving at the edge of the Earth's atmosphere consist of about 50% protons and 25% alpha particles. In the upper atmosphere, **cosmic ray protons collide with the atoms existing there** and decay into positive pions and neutrons {p (uud) → π^+ (ud`) + n (udd)}, and into negative pions and double positive delta baryons {p (uud) → $\pi^-$ (u`d) + Δ^{++} (uuu)}.

The collision of a high energy cosmic ray proton p (uud) with an atom may cause an up quark of the proton to emit a gluon and the up quark remains itself as an up quark u. The gluon then decays into a down quark d and an antidown quark d`. This antidown quark d` combines with that up quark u of proton which had emitted the gluon and produces a **positive pion π⁺ (ud`)** and the down quark d (produced by the decay of the gluon) combines with the remaining up quark and the down quark of the proton and produces a **neutron n (udd).**

Similarly, **the collision of a high energy cosmic ray proton p (uud)** with an atom may cause the down quark of the proton to emit a gluon and the down quark remains itself as a down quark d. The gluon then decays into an up quark u and an antiup quark u`. This antiup quark u` combines with that down quark d of the proton which had emitted the gluon and produces a **negative pion π⁻ (u`d)** and the up quark u (produced by the decay of the gluon) combines with the two up quarks of the proton and produces a **double positive delta baryon Δ⁺⁺ (uuu).** The double positive delta baryon Δ⁺⁺ (uuu) decays into a proton p (uud) and a positive pion π⁺ (ud`) straight at the point, where it is produced because the Δ⁺⁺ (uuu) will travel almost zero distance in its lifetime τ = 5.6×10^{-24} s.

note: The production of the positive pion and the negative pion is not due to the decay of the proton. This is due to the collisions of the protons with the atoms. (mean lifetime of a proton is ~ 10^{33} years).

note: The **double positive delta baryon Δ⁺⁺ (uuu)** was discovered in cloud chamber in 1952 when positive pions were incident on proton target. **π⁺ (ud`) + p (uud) → Δ⁺⁺ (uuu).**

Double positive delta baryon Δ⁺⁺ (uuu) was the first of nearly one hundred resonance states of baryons and mesons discovered.

he **positive pion π⁺ (ud`)** decays into an anti muon μ⁺ and a muon-type neutrino $v_\mu$ after travelling some distance 7.8 meter in its life time $\tau = 2.6 \times 10^{-8}$ s (c τ = 3 × 10⁸ m/s × 2.6 ×10⁻⁸ s = 7.8 m): **π⁺ (ud`)** → μ⁺ + v_μ.

The **negative pion π⁻ (u`d)** decays into a muon μ⁻ and a muon-type antineutrino $v_\mu$` after travelling some distance 7.8 meter in its life time τ = 2.6 ×10⁻⁸ s: **π⁻ (u`d)** → μ⁻ + $v_\mu$`

(Already explained in the topic: Examples of The Decays - examples 1 and 2)

High energy pions survive longer enough to reach the Earth surface.

Neutrons, pions, muons, neutrinos form secondary cosmic rays and move towards the Earth surface which is about 15 km below the upper atmosphere.

The muons are the principle component of cosmic rays at sea level. **The mean lifetime of muon μ⁻ is about 2.2 micro seconds** but if it is traveling with the velocity nearly equal to that of light, then the **velocity of time in the frame of the muon will decrease** by a factor of $1/(1-v^2/c^2)^{1/2}$. That is, the mean lifetime of muon in the frame of muon is always 2.2 micro seconds, but **mean lifetime of the muon in the frame of the Earth is increased** to 2.2 × {1 / $(1-v^2/c^2)^{1/2}$} micro seconds.

The muon has rest energy of 105 MeV. If the total energy of muon in the upper atmosphere is E = 10 GeV = 10000 MeV, then from the Einstein's equation $E = m_0c^2/(1-v^2/c^2)^{1/2}$, $1/(1-v^2/c^2)^{1/2} = E / m_0c^2 = 10000$ MeV/105 MeV = 95.23, hence in this case, the mean lifetime of the muon in the frame of the Earth will be 2.2 × 95.23 = 209.5 micro seconds.

Also, $1/(1-v^2/c^2)^{1/2} = 95.23$ implies the velocity of muons v = .999c or v ~ c = 3 × 10⁸ m/s and a muon moving with the velocity nearly equal to that of light will **travel** 15 kilometers to reach the Earth surface in (15 km)/(3 × 10^8 m/s) = 50 micro seconds whereas its mean lifetime in the frame of the Earth is 209.5 micro seconds (if total energy is 10 GeV), i.e. greater than 50 micro seconds, hence before its decay ($\mu^- \rightarrow e^- + v_e` + v_\mu$ and $\mu^+ \rightarrow e^+ + v_e + v_\mu`$), muon will reach the Earth surface.

note: Some of the positive pions and the negative pions, produced by the collisions of the cosmic rays protons and the atoms in the upper atmosphere, may themselves collide with the high energy cosmic rays protons to produce other particles too, e.g. positive kaons K^+ (us`), negative kaons $K^-$ (u`s), neutral kaons K^0 (ds`), anti-neutral kaons $K^{0`}$ (d`s), lambda baryons Λ^0 (uds), and positive sigma baryons Σ^+ (uus):

π⁻ (u`d) + p (uud) → Λ⁰ (uds) + K⁰ (ds`) (See the topic: Discovery of Strange Quark)
π⁺ (ud`) + p (uud) → Σ⁺ (uus) + K⁺ (us`)
π⁺ (ud`)** + p (uud) → **K⁺ (us`) + K⁰` (d`s) + p (uud)

(To understand, how collisions of **π⁺ (ud`)**, **π⁻ (u`d)** with the **protons** may produce **K⁺ (us`)**, **K⁻ (u`s)**, **Λ⁰ (uds)**, **Σ⁺ (uus)** and other particles, see the topic: Examples of Interactions).

note: Mean lifetimes of K^+ (us`), $K^-$ (u`s), Λ^0 (uds), Σ^+ (uus) are 120 × 10^{-10}s, 120 × 10^{-10}s, 2.6 × 10^{-10} s, 0.8 × 10^{-10} s respectively, and they further decay in the upper atmosphere into positive pions, negative pions and other particles.
Even if they are moving with the velocity nearly equal to that of light, they can travel only some meters (e.g. kaons) or only some centimeters (e.g. lambda baryons, sigma baryons) before the decay.

Discovery of Strange Quark

The <u>strangeness</u> S is a property which causes the slow decay through the W bosons (the mediators of weak interactions) of the particles which have been produced through the strong or the electromagnetic interactions. The strong and the electromagnetic interactions **occur** in 10^{-23} and 10^{-19} seconds respectively, whereas the weak interaction **takes** 10^{-10} seconds, i.e. 10^{13} and 10^9 times more time than the strong and the electromagnetic interactions respectively.

The strange quark was discovered in the bubble chamber exposed to cosmic rays. What actually happened is roughly as follows: When a negative pion π^- (u`d) of cosmic rays entered the bubble chamber, it made a track. After some time, $\pi^-$ (u`d) interacted with one of the protons **p (uud)** of hydrogen atoms in the bubble chamber and produced two neutral particles, i.e. lambda baryon and neutral kaon: π^- (u`d) + p (uud) → $\Lambda^0$ (uds) + $K^0$ (ds`). After this interaction, track disappeared as the negative pion was no more and the neutral particles leave no track.

In the interaction, π^- (u`d) + p (uud) → $\Lambda^0$ (uds) + $K^0$ (ds`), **one up quark of the proton and the antiup quark of the negative pion annihilate to a gluon which** then materializes into a **strange quark s** and an **antistrange quark s`**. This **antistrange quark s`** combines with the down quark of the negative pion to produce a **neutral kaon K^0 (ds`).**

The **strange quark s** (produced by the decay of the gluon) combines with the down quark and the surviving up quark of the proton to produce a **lambda baryon Λ^0 (uds).**

Mean lifetime of lambda baryon Λ^0 (uds) is $\tau \sim 2.6 \times 10^{-10}$ s, and $\tau \sim 2.6 \times 10^{-10}$ s after the interaction of π^- (u`d) and p (uud), somewhere else, a pair of track appeared, when $\Lambda^0$ **(uds) decayed** into a proton p (uud) and a negative pion $\pi^-$ (u`d) which being charged particles formed tracks as they travelled. Thus, Λ^0 **(uds)** was identified through its decay products p (uud) and π^- (u`d). The appearance of this pair of track implied negative pion $\pi^-$ (u`d) and **proton p (uud)** had interacted and had disintegrated into Λ^0 **(uds)** and some other particle. It meant, that the other particle {later called K^0 **(ds`)**} was somewhere else which had not still been seen.

Mean lifetime of neutral kaon K^0 (ds`) and anti neutral kaon anti-$K^0$ (d`s) is $\tau \sim 120 \times 10^{-10}$ s, and $\tau \sim 120 \times 10^{-10}$ s after the interaction of π^- **(u`d)** and **p (uud)**, another pair of track again appeared, when $K^0$ (ds`) **disintegrated** into two charged particles, e.g. into π^+ (ud`) and $\pi^-$ (u`d).

This neutral kaon K^0 (ds`) produced **rapidly** (in $10^{-23}$ s) via strong interaction: $\pi^-$ **(u`d) + p (uud)** $\rightarrow \Lambda^0$ **(uds) +** K^0 **(ds`)**, but decayed much **slowly**. That is, **due to the large mass** of $K^0$ (ds`), it must have been highly unstable and therefore should have decayed **rapidly** through the gluon (the mediators of strong interaction) or photon (the mediator of electromagnetic interactions) but it **decayed slowly** in $\sim 10^{-10}$ s through W boson (the mediator of weak interactions).

This implied neutral kaon K^0 (ds`) **had** a new flavour of the quark called the strange quark which disintegrated much **slowly than expected.** This strange quark was given a new quantum number called the 'strangeness S'.

Massive particles decay rapidly in $\sim 10^{-23}$ seconds but the decay of a massive particle having a strange quark as its constituent {e.g. **Kaons –** K^+ (us`), $K^0$ (ds`), $K^{0`}$ (d`s), K^- (u`s)} into lighter particles having no strange quark {e.g. **Pions** – $\pi^+$(ud`), π^0 (uu`), $\pi^-$ (u`d)} **occurs** in ~ 10^{-10} seconds, i.e. ~ 10^{13} times **slower than expected.** This is so because to decay into the lighter particles having up and down quarks only, the strange quark would have to disappear after the decay and the strangeness quantum number won't be conserved but strangeness is conserved in the strong interactions/decays and electromagnetic interactions/decays, so they **decay slowly** through weak decay, where strangeness is not conserved. It is the strange behaviour of the strange quark, hence it was called the strange quark.

(Why strangeness is conserved in strong and electromagnetic interactions?

The only answer is - Nature behaves this way)

note: Regarding the interaction, π^- (u`d) + p (uud) → $\Lambda^0$ (uds) + $K^0$ (ds`), Λ^0 (uds) having a strange quark also decayed weakly into p (uud) and π^- (u`d) having no strange quark.

note:

Cosmic rays contain very high energy particles which are energetic enough to **penetrate** into the nucleus. Thus cosmic rays became a very useful tool to discover new particles. By using cosmic rays, numbers of new particles were discovered

The positron was discovered in 1932 in cloud chamber placed in a magnetic field and exposed to cosmic rays.

The kaons and the pions were discovered in 1947 in **cosmic rays** which led the discovery of the strange quark.

The pion to muon decay was discovered in 1947 in cosmic rays. ($\pi^- \rightarrow \mu^- + \bar{v}_\mu$ and $\pi^+ \rightarrow \mu^+ + v_\mu$)

The charged pions ($\pi^+ \pi^-$) were first produced in accelerators in 1948.

The neutral pion π^0 was first observed in 1950. Pion decay to 2 photons was also first observed in 1950.

The lambda baryon was first observed in October 1950.

The **double positive delta baryon Δ^{++} (uuu)** was discovered in cloud chamber in 1952, when positive pions were incident on proton target.

$\pi^+ (u\bar{d}) + p (uud) \rightarrow \Delta^{++} (uuu)$.

Same year bubble chamber was invented. The charged sigma baryons and the charged Xi baryons were first observed in the bubble chamber in 1953.

note: Bubble chamber is placed between the plates of a giant magnet. Thus, there is a magnetic field in the bubble chamber. In such a magnetic field of strength B, a particle of electric charge q and momentum p will move in a circle of radius R given by R = pc/qB. The curvature of the track (which gives the value of R) in a known magnetic field B can be used to determine the momentum p of the particle.

In 1952, the first of the modern particle accelerators began operating and soon it was possible to produce particles having strange quarks in the laboratory. Before this, the only source had been cosmic rays. The antiproton was first observed in 1956 in accelerator experiments. The rho, omega and eta mesons were first observed in 1961.

Examples of the Decays

1. The decay of the positive pion π^+ (ud`) into an antimuon μ^+ and a muon-type neutrino v_μ:

π^+ (ud`) $\rightarrow \mu^+ + v_\mu$

In this decay, the up quark u (charge: +2/3e) and the antidown quark d` (charge: +1/3e) of the positive pion annihilate to a W^+ boson (charge: +1e) which then decays into an **antimuon μ^+** and a **muon-type neutrino v_μ.**

2. The decay of the negative pion π^- (u`d) into a muon $\mu^-$ and a muon-type antineutrino $v_\mu$`:

π^- (u`d) $\rightarrow \mu^- + v_\mu$`

In this decay, the antiup quark u` (charge: -2/3e) and the down quark d (charge: -1/3e) of the negative pion annihilate to a $W^-$ boson (charge: -1e) which then decays into a **muon $\mu^-$** and a **muon-type antineutrino $v_\mu$`.**

3. The decay of the muon μ^- into an electron e^-, an electron-type antineutrino v_e` and a muon-type neutrino v_μ:

$\mu^- \rightarrow e^- + v_e` + v_\mu$

In this decay, the muon μ^- with one unit of negative electric charge emits that one unit of negative electric charge by emitting a W^- boson and therefore transforms into the corresponding electrically neutral lepton that is, **muon-type neutrino v_μ.**

The W^- boson subsequently decays into an **electron e^-** and the corresponding antineutrino, i.e. **electron-type antineutrino v_e`.**

note: In this decay, the lepton number is conserved. The muon μ^- having lepton number L = +1 decays into e^- (L= +1), v_e` (L= -1) and v_μ (L= +1).

4. The decay of the lambda baryon Λ^0 (uds) into a proton p (uud) and a negative pion π^- (u`d):

Λ^0 (uds) $\rightarrow$ p (uud) + π^- (u`d)

In this decay, strangeness is not conserved. Strangeness is S = -1 and 0 before and after the decay respectively, thus this decay is through the weak process that is, through a W boson.

In this decay, the **strange quark** (charge: -1/3e) of the lambda baryon **emits a W⁻ boson** and **transforms into** an **up quark u** (charge: +2/3e) and the lambda baryon transforms into a **proton p (uud).**

The W⁺ boson subsequently decays into an **antiup quark u`** (charge: +2/3e) and a **down quark d** (charge: -1/3e) which combine to produce a **negative pion $\pi^-$ (u`d).**

note: Whenever a W boson is involved in an interaction/decay, it is a weak interaction/decay and a **W boson may be absorbed or emitted by a quark because quarks possess weak charges** (in addition to strong and electric charges), hence quarks can involve in the weak interactions/decays through W⁺⁻ and Z⁰ bosons.

5. The decay of the lambda baryon Λ^0 (uds) into a neutron n (udd) and a neutral pion π^0 (uu`):

Λ^0 (uds) $\rightarrow$ n (udd) + π^0 (uu`)

In this decay too, strangeness is not conserved. Strangeness is S = -1 and 0 before and after the decay respectively, thus this decay is through the weak process that is, through a W boson.

In this decay, the **strange quark** (charge: -1/3e) of the lambda baryon emits a W⁻ boson and **transforms into** an **up quark u** (charge: +2/3e). The **W⁻ boson subsequently** decays into an **antiup quark u` (charge: -2/3e)** and a **down quark d (charge: -1/3e). This antiup quark u`** and the **up quark u** (produced by the transformation of the strange quark) combine to produce a **neutral pion π⁰ (uu`).**

The **down quark d** (produced by the decay of the W⁻ boson) combines with the up quark and the down quark of the lambda baryon and produces a **neutron n (udd).**

note: 64% of all lambda baryons decay into p (uud) + π⁻ (u`d) and 36% into n (udd) + π⁰ (uu`)

note: lambda baryon cannot decay via strong process into negative kaon and proton or antineutral kaon and neutron, i.e. decays: Λ^0 (uds) → K⁻ (u`s) + p (uud) and $\Lambda^0$ (uds) → K⁰` (d`s) + n (udd) are forbidden because although, kaons are lightest particles having strange quark but the mass of the lambda baryon (1116 MeV) is less than the mass of the kaon (493 MeV) plus the mass of the nucleon (939 MeV). As in these decays, strangeness is conserved, thus if the mass of the lambda baryon were larger than the sum of the masses of the kaons and nucleons, lambda baryon might emit a gluon which might then decay into u and u` or into d and d` to produce these decay products but these strong decays are kinematically forbidden, thus lambda baryon decays weakly into a nucleon (proton or neutron) and a pion.

6. The decay of the negative Xi baryon Ξ⁻ (dss) into a lambda baryon Λ⁰ (uds) and a negative pion π⁻ (u`d):

Ξ⁻ (dss) → Λ⁰ (uds) + π⁻ (u`d)

In this decay too, strangeness is not conserved. Strangeness is $S = -2$ and -1 before and after the decay respectively, thus this decay is through the weak process.

In this decay, one **strange quark** (charge: -1/3e) of the negative Xi baryon **emits a W⁻ boson** and **transforms into** an **up quark u** (charge: +2/3e) and the negative Xi baryon transforms into a **lambda baryon Λ⁰ (uds).**

The W⁻ boson (charge: -1e) then decays into an **antiup quark u`** (charge: -2/3e) and a **down quark d** (charge: -1/3e) which combine to produce a **negative pion π⁻ (u`d).**

7. The decay of the neutral Xi baryon Ξ⁰ (uss) into a lambda baryon Λ⁰ (uds) and a neutral pion π⁰ (uu`):

Ξ⁰ (uss) → Λ⁰ (uds) + π⁰ (uu`)

In this decay too, strangeness is not conserved. Strangeness is $S = -2$ and -1 before and after the decay respectively, thus this decay is through the weak process.

In this decay, one **strange quark** (charge: -1/3e) of the neutral Xi baryon **emits a W⁻ boson** and **transforms into** an **up quark u** (charge: +2/3e).

The W⁻ boson (charge: -1e) then decays into an **antiup quark u`** (charge: -2/3e) and a **down quark d** (charge: - 1/3e). This **antiup quark u`** combines with the **up quark u** (produced by the transformation of the strange quark) and produces a **neutral pion π⁰ (uu`).**

The **down quark d** (produced by the decay of the W⁻ boson) combines with the up quark and the surviving strange quark of the neutral Xi baryon and produces a **lambda baryon Λ⁰ (uds)**.

8. The decay of the double positive delta baryon Δ⁺⁺ (uuu) into a proton p (uud) and a positive pion π⁺ (ud`):

Δ⁺⁺ (uuu) → p (uud) + π⁺ (ud`)

In this decay, strangeness is conserved. Strangeness is zero before as well as after the decay. Thus, this decay is through the strong process that is, through a gluon.

In this decay, one **up quark** of the double positive delta baryon **emits a gluon** and the up quark remains itself as an **up quark u**.

The **gluon then decays into a down quark d and an antidown quark d`**. This **antidown quark d`** combines with that **up quark u** of the double positive delta baryon which had emitted the gluon and produces a **positive pion π⁺ (ud`).**

The **down quark d** (produced by the decay of the gluon) combines with the remaining two up quarks of the double positive delta baryon and produces a **proton p (uud).**

9. The decay of the positive delta baryon Δ⁺ (uud) into a proton p (uud) and a neutral pion π⁰ (uu`):

Δ⁺ (uud) → p (uud) + π⁰ (uu`)

In this decay too, strangeness is conserved. Thus, this decay is through the strong process.

In this case, one **up quark** of the positive delta baryon **emits a gluon** and remains itself as an **up quark u.**

The **gluon then decays into an up quark u and an antiup quark u`.**

This **antiup quark u`** combines with that **up quark u** of the positive delta baryon which had emitted the gluon and produces a **neutral pion π⁰ (uu`).**

The **up quark u** (produced by the decay of the gluon) combines with the remaining up quark and the down quark of the positive delta baryon and produces a **proton (uud).**

Through the emission of a gluon, **Δ⁺ (1232 MeV)** loses energy and transforms into a proton **p (938.232 MeV) which** contains the same composition of the fundamental particles as **Δ⁺, i.e.** two up quarks and one down quark.

10. The decay of the positive delta baryon Δ⁺ (uud) into a neutron n (udd) and a positive pion π⁺ (ud`):

Δ⁺ (uud) → n (udd) + π⁺ (ud`)

In this decay too, strangeness is conserved. Thus, this decay is through the strong process.

In this decay, one **up quark** of the positive delta baryon **emits a gluon** and remains itself as an **up quark u.**

The **gluon then decays into a down quark d and an antidown quark d`.** This **antidown quark d`** combines with that **up quark u** of the positive delta baryon which had emitted the gluon and produces a **positive pion π⁺ (ud`).**

The **down quark d** (produced by the decay of the gluon) combines with the remaining up quark and the down quark of the positive delta baryon and produces a **neutron n (udd).**

11. The decay of the negative omega baryon Ω⁻ (sss) into a negative Xi baryon Ξ⁻ (dss) and a neutral pion π⁰ (uu`):

Ω⁻ (sss) → Ξ⁻ (dss) + π⁰ (uu`)

This is a weak decay as strangeness is not conserved.

In this decay, **one of the strange quarks** (charge: -1/3e) of the negative omega baryon **emits a W⁻ boson** and **transforms into** an **up quark u** (charge: +2/3e).

The W⁻ boson (charge: -1e) then decays into an **antiup quark u`** (charge: -2/3e) and a **down quark d** (charge: -1/3e). This **antiup quark u`** combines with the **up quark u** (produced by the transformation of the strange quark) and produces a **neutral pion π⁰ (uu`).**

The **down quark d** (produced by the decay of the W⁻ boson) combines with the remaining two strange quarks of the negative omega baryon and produces a **negative Xi baryon Ξ⁻ (dss).**

12. The decay of the omega baryon Ω⁻(sss) into a neutral Xi baryon Ξ⁰ (uss) and a negative pion π⁻(u`d):

Ω⁻(sss) → Ξ⁰ (uss) + π⁻(u`d)

In this decay, strangeness is not conserved. Strangeness is S = -3 and -2 before and after the decay respectively, thus this decay is through the weak process.

In this decay, **one of the strange quarks** (charge: -1/3e) of the omega baryon **emits a W⁻ boson** and **transforms into** an **up quark u** (charge: +2/3e) and the omega baryon transforms into a **neutral Xi baryon Ξ⁰ (uss).**

The W⁻ boson (charge: -1e) then decays into an **antiup quark u`** (charge: -2/3e) and a **down quark d** (charge: - 1/3e) which combine to producie a **negative pion π⁻(u`d).**

13. The decay of the negative omega baryon Ω⁻(sss) into a lambda baryon Λ⁰ (uds) and a negative kaon K⁻(u`s):

Ω⁻(sss) → Λ⁰ (uds) + K⁻(u`s)

This is a weak decay as strangeness is not conserved.

In this decay, **one of the strange quarks** (charge: -1/3e) of the negative omega baryon **emits a W⁻ boson** and **transforms into** an **up quark u** (charge: +2/3e).

The W⁻ boson (charge: -1e) then decays into an **antiup quark u`** (charge: -2/3e) and a **down quark d** (charge: -1/3e). This **antiup quark u`** combines with another strange quark of the negative omega baryon and produces a **negative kaon K⁻(u`s).**

The **down quark d** (produced by the decay of the W⁻ boson) combines with the **up quark** (produced by the transformation of the strange quark) and the remaining strange quark of the negative omega baryon and produces a **lambda baryon Λ⁰ (uds).**

(Above three examples show, how a single particle (in this case, the negative omega baryon) can decay through different modes).

14. The decay of the positive sigma baryon Σ^+ (uus) into a proton p (uud) and a neutral pion π^0 (uu`):

Σ^+ (uus) $\rightarrow$ p (uud) + π^0 (uu`)

In this decay, strangeness is not conserved. Strangeness is S = -1 and 0 before and after the decay respectively, thus this decay is through the weak process that is, through a W boson.

In this case, the **strange quark** (charge: -1/3e) of the positive sigma baryon **emits a W⁻ boson** (and therefore emits charge: -1e) and **transforms into** an **up quark u** (charge: -1/3e +1e = +2/3e).

Then, **one of the two up quarks** (charge: +2/3e) of the positive sigma baryon **absorbs that W⁻ boson** and **transforms into** a **down quark d** (charge: +2/3e -1e = -1/3e).

The second up quark of the positive sigma baryon **emits a gluon** and remains itself as an **up quark u.**

The **gluon subsequently decays into an up quark u and an antiup quark u`.** This **antiup quark u`** combines with that **up quark u** of the positive sigma baryon which had emitted the gluon and produces a **neutral pion π^0 (uu`).**

The **up quark u** (produced by the decay of the gluon) combines with the **up quark u** and the **down quark d** (produced by the exchange of a W⁻ boson between the strange quark and an up quark of the positive sigma baryon) and produces a **proton (uud).**

(This decay of the positive sigma baryon involves the weak process and the strong process both, as a W boson and a gluon both act as mediators in this decay).

Another method for the decay of the positive sigma baryon Σ^+ (uus) into a proton p (uud) and a neutral pion π^0 (uu`): $\Sigma^+$ (uus) → p (uud) + $\pi^0$ (uu`).

In this decay, the **strange quark** (charge: -1/3e) of the positive sigma baryon **emits a W⁻ boson** and **transforms into** an **up quark u** (charge: +2/3e).

The W⁻ boson subsequently decays into an **antiup quark u`** (charge: -2/3e) and a **down quark d** (charge: -1/3e). This **antiup quark u`** combines with the **up quark u** (produced by the transformation of the strange quark) and produces a **neutral pion π^0 (uu`).**

The **down quark d** (produced by the decay of the W⁻ boson) combines with the two up quarks of the positive sigma baryon and produces a **proton p (uud).**

15. The decay of the positive sigma baryon Σ^+ (uus) into a neutron n (udd) and a positive pion π^+ (ud`):

Σ^+ (uus) → n (udd) + π^+ (ud`)

This is a weak decay as strangeness is not conserved.

In this decay, the **strange quark** (charge: -1/3e) of the positive sigma baryon **emits a W⁻ boson** (and therefore emits charge: -1e) and **transforms into** an **up quark u** (charge: -1/3e +1e = +2/3e).

Then **one up quark** (charge: +2/3e) of the positive sigma baryon **absorbs that W⁻ boson** and **transforms into** a **down quark d** (charge: +2/3e -1e = -1/3e).

The second up quark of the positive sigma baryon **emits a gluon** and remains itself as an **up quark u.**

The **gluon subsequently decays into a down quark d and an antidown quark d`.** This **antidown quark d`** combines with that **up quark u** of the positive sigma baryon which had emitted the gluon and produces a **positive pion π^+ (ud`).**

The **down quark d** (produced by the decay of the gluon) combines with the **up quark u** and the **down quark d** (produced by the exchange of a W⁻ boson between the strange quark and an up quark of the positive sigma baryon) and produces a **neutron n (udd).**

(This decay of the positive sigma baryon involves weak process and strong process both, as a W boson and a gluon both act as mediators in this decay)

(The second method as explained in the last example is not possible for the decay of the positive sigma baryon into a neutron and a positive pion (ud`) as the W⁻ boson may decay into down quark d but cannot decay into antidown quark d` necessary to produce a positive pion)

16. The decay of the neutral sigma baryon Σ⁰ (uds) into a lambda baryon Λ⁰ (uds) and a photon γ:

Σ⁰ (uds) → Λ⁰ (uds) + γ (electromagnetic decay)

In this decay, the neutral sigma baryon Σ⁰ (uds) decays through the electromagnetic process without violating conservation of strangeness. Here, up, down or strange quark of the neutral sigma baryon **emits a photon** and remains itself as an **up, down** or **strange quark**.

Through the emission of photon, **Σ⁰ (1192.6 MeV)** loses energy and transforms into **Λ⁰ (1115.6 MeV) which** contains the same composition of the fundamental particles as **Σ⁰, i.e.** one up quark, one down quark and one strange quark.

17. The decay of the negative sigma baryon Σ⁻ (dds) into a neutron n (udd) and a negative pion π⁻ (u`d):

$$\Sigma^- \text{(dds)} \rightarrow n \text{(udd)} + \pi^- \text{(u`d)}$$

In this decay, strangeness is not conserved. Strangeness is S = -1 and 0 before and after the decay respectively, thus this decay is through the weak process that is, through a W boson.

In this decay, the **strange quark** (charge: -1/3e) of the negative sigma baryon **emits a W⁻ boson** and **transforms into** an **up quark u** (charge: +2/3e) and the negative sigma baryon transforms into a **neutron n (udd).**

The W⁻ boson then decays into an **antiup quark u`** (charge: -2/3e) and a **down quark d** (charge: -1/3e) which combine to produce a **negative pion π⁻ (u`d).**

18. The decay of the positive sigma-star baryon Σ*⁺ (uus) into a positive sigma baryon Σ⁺ (uus) and a neutral pion π⁰ (uu`):

$$\Sigma^{*+} \text{(uus)} \rightarrow \Sigma^+ \text{(uus)} + \pi^0 \text{(uu`)}$$

This is a strong decay as strangeness is conserved.

In this decay, one **up quark** of the positive sigma-star baryon **emits a gluon** and remains itself as an **up quark u.**

The **gluon then decays into an up quark u and an antiup quark u`.** This **antiup quark u`** combines with that **up quark u** of the positive sigma-star baryon which had emitted the gluon and produces a **neutral pion π⁰ (uu`).**

The **up quark u** (produced by the decay of the gluon) combines with the remaining up quark and the strange quark of the positive sigma-star baryon to produce a **positive sigma baryon Σ⁺ (uus).**

Through the emission of the gluon, **Σ*⁺ (1382.8 MeV)** loses energy and transforms into **Σ⁺ (1189.3 MeV) which** contains the same composition of the fundamental particles as **Σ*+**, i.e. two up quarks and one strange quark.

19. The decay of the positive sigma-star baryon Σ*+ (uus) into a lambda baryon Λ⁰ (uds) and a positive pion π⁺ (ud`):

Σ*+ (uus) → Λ⁰ (uds) + π⁺ (ud`)

This is a strong decay as strangeness is conserved.

In this decay, one **up quark** of the positive sigma-star baryon **emits a gluon** and remains itself as an **up quark u**.

The **gluon then decays into a down quark d and an antidown quark d`.** This **antidown quark d`** combines with that **up quark u** of the positive sigma-star baryon which had emitted the gluon and produces a **positive pion π⁺ (ud`).**

The **down quark d** (produced by the decay of the gluon) combines with the remaining up quark and the strange quark of the positive sigma-star baryon to produce a **lambda baryon Λ⁰ (uds).**

20. The decay of the negative Xi-star baryon Ξ^{*-} (dss) into a negative Xi baryon Ξ^-(dss) and a neutral pion π^0(dd`):

Ξ^{*-}(dss) $\rightarrow$ Ξ^-(dss) + π^0(dd`)

This is a strong decay as strangeness is conserved.

In this decay, the **down quark** of the negative Xi-star baryon **emits a gluon** and remains itself as a **down quark d**.

The **gluon then decays into a down quark d and an antidown quark d`**. This **antidown quark d`** combines with that **down quark d** of the negative Xi-star baryon which had emitted the gluon and produces a **neutral pion π^0(dd`).**

The **down quark d** (produced by the decay of the gluon) combines with the two strange quarks of the negative Xi-star baryon and produces a **negative Xi baryon Ξ^- (dss).**

Through the emission of a gluon, Ξ^{*-}**(1535.0 MeV)** loses energy and transforms into Ξ^-**(1321.7 MeV) which** contains the same composition of the fundamental particles as Ξ^{*-}, i.e. one down quark and two strange quarks.

note: The **down quark d and the antidown quark d` produced by the decay of the gluon cannot combine to produce a neutral pion** $\pi^0$**(dd`)** because these down and antidown quarks move in the opposite directions. See the note in example 35.

21. The decay of the negative Xi-star baryon Ξ^{*-} (dss) into a neutral Xi baryon Ξ^0(uss) and a negative pion π^-(u`d):

Ξ^{*-}(dss) $\rightarrow$ Ξ^0(uss) + π^-(u`d)

This is a strong decay as strangeness is conserved.

In this decay, the **down quark** of the negative Xi-star baryon **emits a gluon** and remains itself as a **down quark d**.

The **gluon then decays into an up quark u and an antiup quark u`.**

This **antiup quark u`** combines with that **down quark d** of the negative

Xi-star baryon which had emitted the gluon and produces a **negative pion π⁻ (u`d)**.

The **up quark u** (produced by the decay of the gluon) combines with the two strange quarks of the negative Xi-star baryon and produces a **neutral Xi baryon Ξ⁰ (uss)**.

22. The decay of the positive kaon K⁺ (us`) into an antimuon μ⁺ and a muon-type neutrino v_μ:

K⁺ (us`) → μ⁺ + v_μ

In this decay, the up quark u (charge: +2/3e) and the antistrange quark s` (charge: +1/3e) of the positive kaon annihilate to a W⁺ boson (charge: +1e) which then decays into an **antimuon μ⁺** and a **muon-type neutrino v_μ**.

23. The decay of the positive kaon K⁺ (us`) into a neutral pion π⁰ (uu`) and a positive pion π⁺ (ud`):

K⁺ (us`) → π⁰ (uu`) + π⁺ (ud`)

This is a weak decay as strangeness is not conserved.

In this decay, the **antistrange quark** (charge: +1/3e) of the positive kaon **emits a W⁺ boson** and **transforms into** an **antiup quark u`** (charge: -2/3e) and the positive kaon transforms into a **neutral pion π⁰ (uu`)**.

The W⁺ boson then decays into an **up quark u** (charge: +2/3e) and an **antidown quark d`** (charge: +1/3e) which combine to produce a **positive pion π⁺ (ud`)**.

24. The decay of the positive kaon K⁺ (us`) into a neutral pion π⁰ (uu`), a positron e⁺ and an electron-type neutrino v_e.

$K^+ (us`) \rightarrow \pi^0 (uu`) + e^+ + v_e$

This is a weak decay as strangeness is not conserved.

In this decay, the **antistrange quark** (charge: +1/3e) of the positive kaon **emits a W⁺ boson** and **transforms into** an **antiup quark u`** (charge: -2/3e) and the positive koon transforms into a **neutral pion π⁰ (uu`).**

The W⁺ boson subsequently decays into a **positron e⁺** and the corresponding neutrino, i.e. **electron-type neutrino v_e.**

25. The decay of the positive kaon K⁺ (us`) into a neutral pion π⁰ (uu`), an antimuon μ⁺ and a muon-type neutrino v_μ.

$K^+ (us`) \rightarrow \pi^0 (uu`) + \mu^+ + v_\mu$

This is a weak decay as strangeness is not conserved.

In this decay, the **antistrange quark** (charge: +1/3e) of the positive kaon **emits a W⁺ boson** and **transforms into** an **antiup quark u`** (charge: -2/3e) and the positive kaon transforms into a **neutral pion π⁰ (uu`).**

The W⁺ boson subsequently decays into an antimuon **μ⁺** and the corresponding neutrino, i.e. **muon-type neutrino v_μ.**

26. The decay of the positive kaon K⁺ (us`) into three pions – 2 π⁺ (ud`) and 1 π⁻ (u`d):

$K^+ (us`) \rightarrow \pi^+ (ud`) + \pi^+ (ud`) + \pi^- (u`d)$

In this decay, the **antistrange quark** (charge: +1/3e) of the positive kaon **emits a W⁺ boson** and **transforms into** an **antiup quark u`** (charge: -2/3e).

The W⁺ boson then decays into an **up quark u** (charge: +2/3e) and an **antidown quark d`** (charge: +1/3e) which combine to produce a **positive pion π⁺ (ud`)**.

The **up quark** of the positive kaon **emits a gluon** and remains itself as an **up quark u.**

The **gluon then decays into a down quark d and an antidown quark d`**. This **antidown quark d`** combines with that **up quark u** of the positive kaon which had emitted the gluon and produces another **positive pion π⁺ (ud`)**.

The **down quark d** (produced by the decay of the gluon) combines with the **antiup quark u`** (produced by the transformation of the antistrange quark) and produces a **negative pion π⁻ (u`d).**

(This decay of the positive kaon involves weak process and strong process both, as a W boson and a gluon both act as mediators in this decay)

27. The decay of the positive kaon K⁺ (us`) into three pions – 1 π⁺ (ud`) and 2 π⁰ (uu`):

K⁺ (us`) → π⁺ (ud`) + π⁰ (uu`) + π⁰ (uu`)

This decay occurs in the same way as explained in the last example except that the **gluon decays into an up quark u and an antiup quark u`** instead of a down quark d and an antidown quark d`. This is explained below.

In this decay, the **antistrange quark** (charge: +1/3e) of the positive kaon **emits a W⁺ boson** and **transforms into** an **antiup quark u`** (charge: -2/3e).

The W⁺ boson then decays into an **up quark u** (charge: +2/3e) and an **antidown quark d`** (charge: +1/3e) which combine to produce a **positive pion π⁺ (ud`).**

The **up quark** of the positive kaon **emits a gluon** and remains itself as an **up quark u.**

The **gluon then decays into a up quark u and an antiup quark u`.** This **antiup quark u`** combines with that **up quark u** of the positive kaon which had emitted the gluon and produces another **neutral pion π⁰ (uu`).**

The **up quark u** (produced by the decay of the gluon) combines with the **antiup quark u`** (produced by the transformation of the antistrange quark) and produces a **neutral pion π⁰ (uu`).**

note: 63.55% of all positive kaons decay into $\mu^+ + v_\mu$

20.66% of all positive kaons decay into π⁺ (ud`) + π⁰ (uu`)

5.59% of all positive kaons decay into π⁺ (ud`) + π⁺ (ud`) + π⁻ (u`d)

5.07 % of all positive kaons decay into π⁰ (uu`) + e⁺ + v_e

3.35 % of all positive kaons decay into π⁰ (uu`) + $\mu^+ + v_\mu$

1.76 % of all positive kaons decay into π⁺ (ud`) + π⁰ (uu`) + π⁰ (uu`)

28. The decay of the positive rho meson ρ^+ (ud`) into a positive pion $\pi^+$ (ud`) and a neutral pion π^0 (uu`):

ρ^+(ud`) $\rightarrow$ $\pi^+$ (ud`) + π^0 (uu`)

This is a strong decay as strangeness is conserved.

In this decay, the **up quark** of the positive rho meson **emits a gluon** and remains itself as an **up quark u**.

The **gluon then decays into an up quark u and an antiup quark u`.**

This **antiup quark u`** combines with that **up quark u** of the positive rho meson which had emitted the gluon and produces a **neutral pion $\pi^0$ (uu`).**

The **up quark u** (produced by the decay of the gluon) combines with the **antidown quark d`** of the positive rho meson and produces a **positive pion $\pi^+$ (ud`).**

Through the emission of a gluon, ρ^+ **(775.2 MeV)** loses energy and transforms into π^+ **(139.5 MeV) which** contains the same composition of the fundamental particles as ρ^+, **i.e.** one up quark and one antidown quark.

note: up quark u and an antiup quark u` produced by the decay of the gluon cannot combine to produce the neutral pion $\pi^0$ (uu`) because these up and antiup quarks move in the opposite directions. See the note in example 35.

29. The decay of the phi meson φ (ss`) into a positive kaon K⁺ (us`) and a negative kaon K⁻ (u`s):

φ (ss`) → K⁺ (us`) + K⁻ (u`s)

In this decay of the φ meson, strangeness is conserved. Strangeness of φ meson is S= -1 + 1 = 0 and the decay products, the positive kaon K⁺ (us`) has S = +1 and the negative kaon K⁻ (u`s) has S = -1, i.e. total strangeness of the decay products is also zero. Thus, this decay of the φ meson is through the strong process that is, through a gluon.

In this decay, the strange quark s or the antistrange quark s` of the φ meson **emits a gluon** and remains itself as a strange quark s or an antistrange quark s`.

The **gluon then decays into an up quark u and an antiup quark u`.** The **up quark u** combines with the **antistrange quark s`** to produce a **positive kaon K⁺ (us`)** and the **antiup quark u`** combines with the **strange quark s** to produce a **negative kaon K⁻ (u`s).**

Another method for the decay of the phi meson φ (ss`) into a positive kaon K⁺ (us`) and a negative kaon K⁻ (u`s): φ (ss`) → K⁺ (us`) + K⁻ (u`s)

The strange quark of the φ meson **emits a gluon** and remains itself as a **strange quark s.**

The antistrange quark of the φ meson also **emits a gluon** and remains itself as an **antistrange quark s`.**

The gluon emitted by the strange quark s decays into a real **antiup quark u`** and a virtual up quark. This real **antiup quark u`** combines with this **strange quark s** to produce a **negative kaon K⁻ (u`s).** The gluon emitted by the antistrange quark s` absorbs this virtual up quark and transforms into a real **up quark u.** This real **up quark u** combines with this **antistrange quark s`** to produce a **positive kaon K⁺ (us`).**

(Thus, in this decay, the phi meson radiates two gluons and there is an exchange of a virtual up quark between these gluons).

30. The decay of the phi meson φ (ss`) into a neutral kaon K⁰ (ds`) and antineutral kaon K⁰` (d`s):

φ (ss`) → K⁰ (ds`) + K⁰` (d`s)

Even in this decay of the φ meson, strangeness is conserved. Thus, this decay of the φ meson is through the strong process.

In this decay, the strange quark s or the antistrange quark s` of the φ meson **emits a gluon** and remains itself as a strange quark s or an antistrange quark s`.

The **gluon then decays into a down quark d and an antidown quark d`**. The **down quark d** combines with the **antistrange quark s`** to produce a **neutral kaon K⁰ (ds`)** and the **antidown quark d`** combines with the **strange quark s** to produce an **antineutral kaon K⁰` (d`s).**

note: This decay may also occur by another method given in the last example.

31: The decay of the phi meson φ (ss`) into a positive pion π⁺ (ud`), a negative pion π⁻ (u`d) and a neutral pion π⁰ (uu`):

φ (ss`) → π⁺ (ud`) + π⁻ (u`d) + π⁰ (uu`)

note: Phi meson decays much more often into two kaons than into three pions in spite of the fact that the three-pion decay is energetically favoured (the mass of the two kaons is about 2*493 MeV or 984 MeV, whereas the mass of the three pions is about 134 MeV + 2*139 MeV or 412 MeV).

**The decay of φ (ss`) into 3π is OZI suppressed but not absolutely forbidden** for the decay φ (ss`) → 3 π does in fact occur, but far less likely than one would have supposed.

In the decay φ (ss`) → π⁺ (ud`) + π⁻ (u`d) + π⁰ (uu`), strangeness is conserved, thus this decay is through the strong process.

In this decay, both the strange quark and the antistrange quark of the φ meson **emit** one gluon each and then the **annihilation** of the strange quark and the antistrange quark of the φ meson produces third gluon. One of these three gluons decays into up and antiup quark, second gluon also decays into up and antiup quark and third gluon decays into down and antidown quark. Thus, three gluons produce two up quarks u, two antiup quarks u`, one down quark d and one antidown quark d`. These quarks and antiquarks combine to produce three pions.

Now, note that in this decay, both the strange quark and the antistrange quark of the phi meson **disappear** that is, the **entire phi meson is transformed into three gluons** and these three **gluons** carry the energy necessary to produce three pions that is, these **are high energy** or hard gluons but **gluons couple weakly at high energies, i.e.** the strength with which they decay into up and antiup quarks or down and antidown quarks decreases, hence this decay is suppressed and is called as **OZI suppressed**.

In OZI-allowed processes, the gluons are typically 'soft' (low energy) and in this regime, the coupling of the gluons to the quarks is strong.

32. The decay of the phi meson φ (ss`) into a positive rho meson ρ⁺ (ud`) and a negative pion π⁻ (u`d):

φ (ss`) → ρ⁺ (ud`) + π⁻ (u`d)

Even in this decay of the φ meson, strangeness is conserved.

In this decay, the strange quark s or the antistrange quark s` of the φ mesons **emits a gluon.** The strange quark s and the antistrange quark s` then **annihilate to another gluon**. One gluon decays into an **up quark u** and an **antiup quark u`** and the other gluon decays into a **down quark d** and an **antidown quark d`**. The **up quark u** combines with the **antidown quark d`** to produce a positive **rho meson ρ⁺ (ud`)** and the **down quark d** combines with the **antiup quark u`** to produce a **negative pion π⁻ (u`d).**

note: Even this decay is OZI suppressed as **entire phi meson φ (ss`) converts into two gluons which** subsequently produce ρ⁺ (ud`) and π⁻ (u`d), i.e. these gluons are highly energetic.

note: 48.9 % of all phi mesons decay into **K⁺ (us`) + K⁻ (u`s)**
34.2 % of all phi mesons decay into **K⁰ (ds`) + K⁰` (d`s)**
15.32 % of all phi mesons decay into **π⁺ (ud`) + π⁻ (u`d) + π⁰ (uu`)** or ρ⁺ (ud`) + π⁻ (u`d)

33. The decay of the neutral K-star meson K*⁰ (ds`) into a neutral kaon K⁰ (ds`) and a neutral pion π⁰ (dd`):

K*⁰ (ds`) → K⁰ (ds`) + π⁰ (dd`)

This is a strong decay as strangeness is conserved.

In this decay, the **antistrange quark** of the neutral K-star meson **emits a gluon** and remains itself as an **antistrange quark s`**.

The **gluon then decays into a down quark d and an antidown quark d`**. This **down quark d** combines with that **antistrange quark s`** of the neutral K-star meson which had emitted the gluon and produces a **neutral kaon K⁰ (ds`).**

The **antidown quark d`** (produced by the decay of the gluon) combines with the **down quark d** of the neutral K-star meson to produce **a neutral pion π⁰ (dd`).**

Through the emission of a gluon, **K*⁰ (895.8 MeV)** loses energy and transforms into **K⁰ (497.6 MeV) which** contains the same composition of the fundamental particles as **K*⁰, i.e.** one down quark and one antistrange quark.

note: Instead of the antistrange quark, the down quark of the neutral K-star meson may also emit a gluon.

34. The decay of the neutral K-star meson K*⁰ (ds`) into a positive kaon K⁺ (us`) and a negative pion π⁻ (u`d):

K*⁰ (ds`) → K⁺ (us`) + π⁻ (u`d)

This is a strong decay as strangeness is conserved.

In this decay, the **down quark** of the neutral K-star meson **emits a gluon** and remains itself as a **down quark d.**

The **gluon then decays into an up quark u and an antiup quark u`.** This **antiup quark u`** combines with that **down quark d** of the neutral K-star meson which had emitted the gluon and produces a **negative pion π⁻ (u`d).**

The **up quark u** (produced by the decay of the gluon) combines with the **antistrange quark s`** of the neutral K-star meson and produces a **positive kaon K⁺ (us`).**

note: Instead of the down quark, the antistrange quark of the neutral K-star meson may also emit a gluon.

35. The decay of the negative K-star meson K*⁻ (u`s) into a negative kaon K⁻ (u`s) and a neutral pion π⁰ (uu`):

K*⁻ (u`s) → K⁻ (u`s) + π⁰ (uu`)

This is a strong decay as strangeness is conserved.

In this decay, the **antiup quark** of the negative K-star meson **emits a gluon** and remains itself as an **antiup quark u`.**

The **gluon then decays into an up quark u and an antiup quark u`.**

This **up quark u** combines with that **antiup quark u`** of the negative K-star meson which had emitted the gluon and produces a **neutral pion π⁰ (uu`).**

The **antiup quark u`** (produced by the decay of the gluon) combines with the strange quark of the negative K-star meson and produces a **negative kaon K⁻ (u`s)**

Through the emission of a gluon, **K*⁻ (891.6 MeV)** loses energy and transforms into **K⁻ (493.6 MeV) which** contains the same composition of fundamental particles as **K*⁻, i.e.** one antiup quark and one strange quark.

note: Instead of the antiup quark, the strange quark of the negative K-star meson may also emit a gluon.

note: As the separation between antiup quark u` and strange quark s (moving in opposite directions) of negative K-star meson K\*⁻ (u`s) increases, a gluon is emitted. However, the **up quark u and the antiup quark u` produced by the decay of the gluon cannot combine to**

produce the neutral pion π^0 (uu`) because these up and antiup quarks too move in the opposite directions. The **strange quark s** of negative K-star meson and the **antiup quark u`** produced by the decay of the gluon move in the same direction and combine to produce a **negative kaon K⁻ (u`s)**. The **antiup quark u`** of the negative K-star meson and the **up quark u** produced by the decay of the gluon move in the same direction with respect to each other but at 180^0 with respect to those quark-antiquark which combine to produce negative kaon K⁻ (u`s). These **antiup quark u` and up quark u** combine to produce **neutral pion π^0 (uu`)**. This means, **the neutral pion $\pi^0$ (uu`) moves at 180^0 with respect to the negative kaon K⁻ (u`s).**

36. The decay of the negative K-star meson K*⁻ (u`s) into an antineutral kaon K⁰` (d`s) and a negative pion $\pi^-$ (u`d):

K*⁻ (u`s) → K⁰` (d`s) + $\pi^-$ (u`d)

This is a strong decay as strangeness is conserved.

In this decay, the **antiup quark** of the negative K-star meson **emits a gluon** and remains itself as an **antiup quark u`.**

The **gluon then decays into a down quark d and an antidown quark d`.** This **down quark d** combines with that **antiup quark u`** of the negative K-star meson which had emitted the gluon and produces a **negative pion π^- (u`d).**

The **antidown quark d`** (produced by the decay of the gluon) combines with the **strange quark s** of the negative K-star meson and produces an **antineutral kaon K⁰` (d`s).**

note: Instead of the antiup quark, the strange quark of the negative K-star meson may also emit a gluon.

37. The decay of the muon μ^- into two electrons e^-, a positron e^+, an electron-type antineutrino $v_e\grave{}$ and a muon-type neutrino v_μ:

$\mu^- \to e^- + e^- + e^+ + v_e\grave{} + v_\mu$

In this decay, the muon μ^- emits a W^- boson and transforms into a **muon-type neutrino v_μ.**

The W^- boson subsequently decays into an **electron e^-** and the corresponding antineutrino, i.e. **electron-type antineutrino $v_e\grave{}$.**

The **electron e^-** then emits a photon which then decays into an **electron e^-** and a **positron e^+.**

note:

The decays: π^+ (ud$\grave{}$) $\to \mu^+ + v_\mu$, π^- (u$\grave{}$d) $\to \mu^- + v_\mu\grave{}$, $\mu^- \to e^- + v_e\grave{} + v_\mu$, K^+ (us$\grave{}$) $\to \mu^+ + v_\mu$ are **leptonic decays** as all the decay products in each of these decays are leptons/antileptons.

The decays: Λ^0 (uds) $\to$ p (uud) + π^- (u$\grave{}$d), Ω^- (sss) $\to \Xi^0$ (uss) + π^- (u$\grave{}$d), K^+ (us$\grave{}$) $\to \pi^+$ (ud$\grave{}$) + π^0 (uu$\grave{}$) are **non-leptonic decays** as none of the decay products in each of these decays are leptons/antileptons.

The decays: Σ^- (dds) $\to$ n (udd) + $e^- + v_e\grave{}$, n (udd) $\to$ p (uud) + $e^- + v_e\grave{}$, K^+ (us$\grave{}$) $\to \pi^0$ (uu$\grave{}$) + $e^+ + v_e$ are **semi-leptonic decays** as some of the decay products in each of these decays are leptons/antileptons.

Parity

The mirror image, i.e. the image in the mirror of any physical process also represents a perfectly possible physical process. The strong and the electromagnetic processes follow mirror symmetry or 'parity invariance', however the weak processes do not. In 1956, an experiment was conducted, in which cobalt 60 nuclei were aligned so that their **spins pointed** in the same direction, say, **along the +z direction or upward.** In such an experiment, cobalt 60 undergoes **beta decay** ($^{60}Co \rightarrow {}^{60}Ni + e^- + v_e`$, i.e. a neutron in the nucleus of Cobalt transforms into a proton and an electron and an electron-type antineutrino are emitted). The electron is emitted in the direction opposite to the direction of the nuclear spin, i.e. **electron is emitted along the –z direction or downward.**

Now, if we examine the mirror image of that same process (assuming the plane of the mirror is parallel to the z–axis i.e perpendicular to the x-y plane), then the image of the cobalt nucleus **in the mirror** will rotate in the opposite direction, hence, in the mirror, **the spin** (which is due to the rotation of a particle on its axis) **of the cobalt nucleus will point along the –z direction, i.e.** downward. However, the direction of the motion of the electron in the mirror will be the same, i.e. along the –z direction, i.e. downward. (**note:** If a particular thing is moving downward, its image in the mirror will also be moving downward). Thus, whereas in real, electron emits in the direction opposite to that of the nuclear spin (the electron is moving along the -z axis and the cobalt nuclear spin is also along the +z axis), **in the mirror, the electron is seen to be emitted in the direction of the nuclear spin** (the electron is seen to be moving along the -z axis and the cobalt nuclear spin is also along the -z axis).

However, this physical **process i.e.** the emission of the electron in the direction of the cobalt nuclear spin **does not occur in nature.** The electron is always emitted in the direction opposite to the direction of the cobalt nuclear spin that is the mirror image of this physical process does not occur. This means, **the parity is not conserved in the weak interaction/decay i.e. the parity is not an invariance of the weak interaction/decay** (β-decay is a weak decay).

As a natural choice, **the direction of the motion is taken along the +z-axis.**

The value of m_s/s along this axis is called the helicity of the particle. Thus, a particle of spin 1/2 (s = 1/2) can have the helicity as +1 (for m_s = 1/2, when spin points upward, i.e. along the +z axis) or -1 (for m_s = -1/2, when spin points downward, i.e. along –z axis).

'+1' is called the **right-handed helicity RH** and '-1' is called the **left-handed helicity LH**.

A particle has **right-handed helicity RH**, if the spin and the velocity of the particle are in the **same** direction, i.e. along the +z axis.

A particle has **left-handed helicity LH**, if the spin and the velocity of the particle are in the **opposite** directions, i.e. spin along the –z axis and velocity along the +z axis).

Suppose, a right-handed electron is moving towards the East with the velocity v, hence its spin too is towards the East. Suppose, someone else is in an inertial system which is also moving towards the East but at a velocity v` > v. For such an observer, in such a frame, he will be at rest and the electron will be moving towards the West with velocity v` - v > 0. Even for that observer (unlike mirror image), the electron will still be spinning the same way, hence the spin will still be towards the East. So this observer will say, it is a left-handed electron (as the motion of the electron is towards the West and its spin is towards the East). Thus, we can convert a right-handed electron into a left-handed one by changing our frame of reference.

But in case of neutrino which is massless and therefore always travels at the speed of light, there can be no observer travelling faster and therefore it is impossible to reverse the 'direction of motion' of a massless neutrino by getting into a faster-moving reference system and therefore the helicity of a neutrino (or any other massless particle) is a fixed and fundamental property.

It was observed that the neutrinos are left-handed (LH), i.e. the spin and the velocity of the neutrino are in the opposite directions and the antineutrinos are right-handed (RH) i.e the spin and velocity of the antineutrino are in the same direction.

In the decay of neutral pion: π^0 (uu`) $\rightarrow \gamma + \gamma$, the decay products i.e. two photons must have the same helicity as explained below.
If the neutral pion π^0 is at rest, then its decay products i.e. the two photons will move in opposite directions (to preserve momentum) after their production. So **if the photon 1 is moving along the +z axis, then the photon 2 will be moving along the -z axis.** Moreover, **since the neutral pion** (a member of the pseudoscalar meson nonet) **has spin 0,** the spins of those two photons must be oppositely aligned so that the total spin of the decay products is zero too. Thus, **if the spin of the photon 1** (moving along the +z axis) **is along the +z axis** i.e if the photon 1 is a RH photon (the spin and the direction of the motion both along -z axis, i.e. parallel), then **the spin of the photon 2** (moving along the -z axis) **will be along the -z axis, i.e.** photon 2 will also be RH.

note:

If RH photon is moving along the +z axis, its spin will also be along the +z axis. Now, if we see the image of this photon in a mirror, whose plane is placed parallel to the z-axis, then the image of the RH photon **in the mirror** will rotate in the opposite direction, hence, in the mirror, **the spin of the photon will point along the –z direction**. However, the direction of the motion of the photon in the mirror will be the same, i.e. along +z direction, i.e. image of RH photon will be LH photon. That is, if the parity operation is applied to RH photon that is, if the mirror image of RH photon is taken, it will become LH photon which exists in nature. Similarly, if the parity operation is applied to LH photon, it will become RH photon which also exists in nature.

The neutral pion decay is an electromagnetic decay which respects parity, i.e. RH photon and its mirror image i.e. LH photon both exist.

Thus, on the average, we get just as many RH photon pairs as LH photon pairs. If in an experiment, 1000 neutral pions decay into photon pairs, then 500 photon pairs will be RH and remaining 500 will be LH.

The decay of the negative pion into the electron and the electron-type antineutrino is highly suppressed.

The **branching ratio** for the decay of the negative pion π^- ($u`d$) into an electron and an electron-type antineutrino $v_e$: $\pi^-$ **($u`d$) $\rightarrow$ e^- + $v_e`$** is 1.23×10^{-4} or 0.0123 %.

That is, for each 1 million (1000000) π^- ($u`d$) produced, 123 will decay into $e^-$ + $v_e`$.

That is, for each 1000000 / 123 or 8130 π^- ($u`d$) produced, 1 will decay into $e^-$ + $v_e`$. All others decay into μ^- + $v_\mu`$.

In the decay of negative pion: $\pi^-(u\grave{}d) \rightarrow e^- + v_e\grave{}$, the decay products i.e. electron and electron-type antineutrino must have the same helicity as explained below.

If the negative pion π^- is at rest, then its decay products i.e. electron and electron-type antineutrino will move in opposite directions (to preserve momentum) after their production. So **if the electron is moving along the +z axis, then the electron-type antineutrino will be moving along the -z axis.** Moreover, **since the negative pion** (a member of the pseudoscalar meson nonet) **has spin 0,** the spins of electron and electron-type antineutrino must be oppositely aligned so that the total spin of the decay products is zero too. Thus, **if the spin of the electron** (moving along the +z axis) **is along the +z axis** i.e if the electron is a RH electron (the spin and the direction of the motion both along +z axis, i.e. parallel), then **the spin of the electron-type antineutrino** (moving along the -z axis) **will be along the -z axis, i.e.** electron-type antineutrino will also be RH.

<u>That is, in the decay $\pi^- \rightarrow e^- + v_e\grave{}$, if electron-type antineutrino is RH, then electron will also be RH.</u>

Now note that, electron-type antineutrino being antineutrino is always RH. However, if the electron were truly massless, then like the neutrino, it would only exist as LH and then the decay $\pi^- \rightarrow e^- + v_e\grave{}$ could never occur at all and since physical electron is very close to being massless, this decay is highly suppressed.

According to the Quantum Field Theory, the parity of a fermion (spin 1/2) must be opposite to that of the corresponding antiparticle while the parity of a boson is the same as its antiparticle.

The **quarks are assumed to have positive or even intrinsic parity (P = +1),** so the antiquarks have negative or odd intrinsic parity (P = -1).

The parity of a composite particle in its ground state (i.e. S state, for which orbital angular momentum L = 0) is the product of the parities of its constituents, i.e. **parity is a multiplicative quantum number,** whereas the charge, strangeness, lepton number, etc. are additive. Thus, all the **baryons** in the baryon octet and baryon decuplet (each consisting of 3 quarks) **have positive or even parity** (P = +1 × +1 × +1 = +1)

All the **mesons** in the pseudoscalar meson nonet and vector meson nonet (each consisting of one quark and one antiquark) **have negative or odd parity** (P = +1 × -1= -1).

Based on the above convention, the **photon** is found to have **negative or odd parity** (P = -1).

<u>A weak interaction/decay can change parity but it does not have to change it necessarily.</u>

For example, **positive kaon K⁺ (us`) may decay into two pions and into three pions.**

K⁺ (us`) → π⁺ (ud`) + π⁰ (uu`) and  K⁺ (us`) → π⁺ (ud`) + π⁺ (ud`) + π⁻ (u`d)

Positive kaon and each of the three pions (π⁺, π⁻, π⁰) are pseudoscalar mesons which have odd parity (P = -1).

Thus, in the decay: **K⁺ (us`) → π⁺ (ud`) + π⁰ (uu`),** the parity of the decay products (π⁺, π⁰) is P = -1× -1 = +1, whereas the parity of the positive kaon is P = -1. Thus, **parity is not conserved in this decay** and this is a weak decay (See the topic: Examples of the Decays – example 23).

In the decay: **K⁺ (us`) → π⁺ (ud`) + π⁺ (ud`) + π⁻ (u`d),** the parity of the decay products (π⁺, π⁻, π⁰) is P = (-1)³ = -1, the same as the parity of the positive kaon. Thus, the **parity is conserved in this decay** but still

this is a weak decay as strangeness is not conserved in this decay. (See the topic: Examples of the Decays - example 26).

note:

A **pseudoscalar meson** is a meson with total spin 0 and odd parity $(J^P=0^-)$.

The pseudoscalar mesons (e.g. kaons, pions) have quark and antiquark spins anti-aligned (S = +1/2 - 1/2) and zero orbital angular momentum (L = 0), resulting in total spin J = L + S = 0.

A **scalar meson** is a meson with total spin 0 and even parity $(J^P=0^+)$. The scalar mesons have quark and antiquark spins anti-aligned (S = +1/2 - 1/2) and zero orbital angular momentum (L = 0), resulting in total spin J = L + S = 0.

A **vector meson** is a meson with total spin 1 and odd parity $(J^P = 1^-)$. The vector mesons (e.g. K-star mesons, rho mesons) have quark and antiquark spins aligned (S = +1/2 + 1/2) and zero orbital angular momentum (L = 0), resulting in total spin J = L + S = 1.

Charge conjugation

The operation of the charge conjugation to a particle converts it into its antiparticle.

Photons's charge conjugation number is C = -1.

A system consisting of spin 1/2 particle and its antiparticle, in a configuration with orbital angular momentum L and total spin S, has charge conjugation number $C = (-1)^{L+S}$.

For **pseudoscalar mesons**, L= 0 and S= 0, thus charge conjugation number $C = (-1)^{0+0} = +1$

For **vector mesons**, L= 0 and S= 1, thus charge conjugation number $C = (-1)^{0+1} = -1$.

Only those particles which are their own antiparticles can have charge conjugation number.

Out of 9 **pseudoscalar mesons**, only K^0, π^0, η, η', $K^{0\grave{}}$ have charge conjugation number (**C = +1**).

Out of 9 **vector mesons**, only K^{*0}, ρ^0, ω, φ, $K^{*0\grave{}}$ have charge conjugation number (**C = -1**)

Charge conjugation is a multiplicative quantum number, and like parity, it is conserved in the strong and electromagnetic interactions.

The **omega meson decays into a neutral pion and a photon: ω →** π^0 **+ γ. This is electromagnetic decay.**

ω (vector meson) has C = -1; the decay products, π^0 (pseudoscalar meson) has C = +1 and γ has C = -1

Thus, in the decay: ω → π^0 + γ, the total charge conjugation number of the decay products is C = +1 × -1 = -1, the same as that of the omega meson. Thus, in this electromagnetic decay, the charge conjugation number is conserved.

The electromagnetic decay $\omega \to \pi^0 + 2\gamma$ is forbidden.

In the decay: $\omega \to \pi^0 + 2\gamma$, the total charge conjugation number of the decay products is $C = +1 \times -1 \times -1 = +1$, whereas the charge conjugation number of the omega meson is $C = -1$. Thus, the charge conjugation number is not conserved in this decay.

The neutral pion decays into two photons: $\pi^0 (u\bar{u}) \to \gamma + \gamma$. This is electromagnetic decay.

π^0 (pseudoscalar meson) has $C = +1$; For n photons, $C = (-1)^n$.

Thus, in the decay: $\pi^0 (u\bar{u}) \to \gamma + \gamma$, the total charge conjugation number of the decay products i.e. 2 photons is $C = (-1)^2 = +1$, the same as the charge conjugation number of the neutral pion. Thus, in this electromagnetic decay, the charge conjugation number is conserved.

The **electromagnetic decay $\pi^0 (u\bar{u}) \to \gamma + \gamma + \gamma$ is forbidden**.

In the decay: $\pi^0 (u\bar{u}) \to \gamma + \gamma + \gamma$, the total charge conjugation number of the decay products i.e. 3 photons is $C = (-1)^3 = -1$, whereas the charge conjugation number of the neutral pion is $C = +1$. Thus, the charge conjugation number is not conserved in this decay.

The **strong decay $\rho^0 (u\bar{u}) \to \pi^0 (u\bar{u}) + \pi^0 (u\bar{u})$ is forbidden.**

ρ^0 (vector meson) has $C = -1$; π^0 (pseudoscalar meson) has $C = +1$.

Thus, in the decay: $\rho^0 (u\bar{u}) \to \pi^0 (u\bar{u}) + \pi^0 (u\bar{u})$, the total charge conjugation number of the decay products is $C = +1 \times +1 = +1$, whereas the charge conjugation number of the rho meson is $C = -1$. Thus, the charge conjugation number is not conserved in this decay.

Thus, the neutral rho meson $\rho^0 (u\bar{u})$ can not emit a gluon to produce up and antiup quark pair which would combine with up and antiup quarks of the rho meson to produce two neutral pions.

Charge conjugation is not a symmetry of the weak interactions/decays. **If charge conjugation operation is applied to a neutrino v (which is always LH), the sign of the weak charge of the neutrino is reversed, that is, it will become LH antineutrino v` which does not exist in nature.** So charged-conjugated version of any process involving neutrinos is a physical process that does not occur in nature. **So this is C-violation.** Purely hadronic weak interactions/decays also show violations of C as well as P.

The CP Operation

To understand the CP operation, consider the decay of the positive pion π^+ (ud`) into an antimuon μ^+ and a muon-type neutrino ν_μ:

π^+ (ud`) $\rightarrow \mu^+ + \nu_\mu$

In this decay, the antimuon μ^+ emitted is always LH. Thus, the decay process can be written as: $\pi^+ \rightarrow \mu^+_{LH} + \nu_\mu$

If we apply **parity operation** to the decay: $\pi^+ \rightarrow \mu^+_{LH} + \nu_\mu$, the decay process will become $\pi^+ \rightarrow \mu^+_{RH} + \nu_\mu$

This can be written as: P $|\pi^+ \rightarrow \mu^+_{LH} + \nu_\mu > = \pi^+ \rightarrow \mu^+_{RH} + \nu_\mu$.

The decay process: $\pi^+ \rightarrow \mu^+_{RH} + \nu_\mu$ does not occur in nature. So this is P-violation.

That is, the weak interactions/decays are not invariant under the parity transformation.

If there were no P-violation, i.e. the weak interactions/decays were invariant under the parity operation, LH antimuons and RH antimuons would be produced in equal amounts in this decay of the positive pions – just as we get just as many RH photon pairs as LH photon pairs in the decay of the neutral pions: π^0 (uu`) $\rightarrow \gamma + \gamma$.

If we apply **charge conjugation operation** to the decay: $\pi^+ \rightarrow \mu^+_{LH} + \nu_\mu$, the decay process will become $\pi^- \rightarrow \mu^-_{LH} + \nu_\mu$

This can be written as: C $|\pi^+ \rightarrow \mu^+_{LH} + \nu_\mu > = \pi^- \rightarrow \mu^-_{LH} + \nu_\mu$

The decay process: $\pi^- \rightarrow \mu^-_{LH} + \nu_\mu$ also does not occur in nature. So this is C-violation.

That is, the weak interactions/decays are not invariant under the charge conjugation operation.

However, if we apply P operation, and then C operation to the decay: $\pi^+ \rightarrow \mu^+_{LH} + v_\mu$, the decay process will become $\pi^- \rightarrow \mu^-_{RH} + \bar{v}_\mu$ as follows:

$P \,|\pi^+ \rightarrow \mu^+_{LH} + v_\mu > \, = \pi^+ \rightarrow \mu^+_{RH} + v_\mu$

$C \,|\pi^+ \rightarrow \mu^+_{RH} + v_\mu > \, = \pi^- \rightarrow \mu^-_{RH} + \bar{v}_\mu$

i.e. CP $|\pi^+ \rightarrow \mu^+_{LH} + v_\mu > \, = \pi^- \rightarrow \mu^-_{RH} + \bar{v}_\mu$

That is, a left-handed antimuon μ^+_{LH} turns into a right-handed muon μ^-_{RH} under combined CP operation

The decay process: $\pi^- \rightarrow \mu^-_{RH} + \bar{v}_\mu$ occurs in nature, i.e. in the decay of the negative pion: $\pi^- \rightarrow \mu^- + \bar{v}_\mu$, the muon μ^- emitted is always RH.

That is, the weak interactions are invariant under combined CP operation.

note: in the above process, now consider the helicity of the neutrino too which is always LH. The P operation on this LH muon-type neutrino v_μ $_{LH}$ will transform it into a RH muon-type neutrino $v_{\mu\,RH}$: $P|\, v_{\mu\,LH} > \, = v_{\mu\,RH}$ **which does not exist** in nature but then the C operation on this RH muon-type neutrino $v_{\mu\,RH}$ will transform it into a RH muon-type antineutrino $\bar{v}_{\mu\,RH}$: $C \,| \, v_{\mu\,RH} > \, = \bar{v}_{\mu\,RH}$, i.e. **CP** $|\, v_{\mu\,LH} > \, = \bar{v}_{\mu\,RH,}$ **which exists in nature**.

So if we include the helicity of the neutrino too in the decay: $\pi^+ \rightarrow \mu^+ + v_\mu$, **CP** $|\pi^+ \rightarrow \mu^+_{LH} + v_\mu > \, = \pi^- \rightarrow \mu^-_{RH} + \bar{v}_\mu$ may be written as **CP** $|\pi^+ \rightarrow \mu^+_{LH} + v_{\mu\,LH} > \, = \pi^- \rightarrow \mu^-_{RH} + \bar{v}_{\mu\,RH}$.

Both these decay processes $\pi^+ \rightarrow \mu^+_{LH} + v_{\mu\,LH}$ and $\pi^- \rightarrow \mu^-_{RH} + \bar{v}_{\mu\,RH}$ occur in nature.

The Branching Ratio

The branching ratios for the decay of the W boson

The **probability or the branching ratio** for the decay of the W⁺ boson into a positron e⁺ and an electron-type neutrino v_e: **W⁺ → e⁺ + v_e** is **0.107 or 10.7 %.**

The **branching ratio** for the decay: **W⁺ → μ⁺ + v_μ** is 0 .106 or 10.6 %.

The **branching ratio** for the decay: **W⁺ → τ⁺ + v_τ** is **0.113 or 11.3 %.**

The **branching ratio** for the decay of the W⁺ boson into hadrons: **W⁺ → hadrons** is **0.674 or 67.4 %.**

That is, for each 1000 W⁺ bosons produced, 107 will decay into e⁺ + v_e, 106 into μ⁺ + v_μ , 113 into τ⁺ + v_τ, and 674 into hadrons.

Similarly, for each 1000 W⁻ bosons produced, 107 will decay into e⁻ + $\bar{v_e}$, 106 into μ⁻ + $\bar{v_\mu}$, 113 into τ⁻ + $\bar{v_\tau}$, and 674 into hadrons.

The branching ratios for the decay of the Z⁰ boson

The **branching ratio** for the decay of the Z⁰ boson into a positron e⁺ and an electron e⁻: **Z⁰ → e⁺ + e⁻** is **0.0336 or 3.36 %.**

The **branching ratio** for the decay: **Z⁰ → μ⁺ + μ⁻** is **0.0336 or 3.36 %.**

The **branching ratio** for the decay: **Z⁰ → τ⁺ + τ⁻** is **0.0337 or 3.37 %.**

The **branching ratio** for the decay of the Z⁰ boson into charm and anticharm quark pair is: **Z⁰ → c$\bar{c}$** is **0.12 or 12 %.**

Similarly, **Br (Z⁰ → b$\bar{b}$) = 0.151 or 15.1 %.**

The quark-antiquark pair, in which a Z⁰ boson decays, may further decay into other hadrons.

The total branching ratio for the decay of the Z⁰ boson into hadrons: **Z⁰ → hadrons** is **0.699 or 69.9 %.**

Remaining 20% branching ratio for the decay of the Z^0 boson is invisible.

That is, for each 10000 Z^0 bosons produced, 336 will decay into e^+ + e^-, 336 into μ^+ + μ^-, 337 into τ^+ + τ^-, and 6990 into hadrons

The branching ratios for some other processes

The **branching ratio** for the decay of the positive sigma baryon Σ^+ (uus) into a proton p (uud) and a neutral pion π^0 (uu`): **$\Sigma^+$ (uus) $\rightarrow$ p (uud) + $\pi^0$ (uu`)** is **0.5157 or 51.57 %.**

The **branching ratio** for the decay: **Σ^+ (uus) $\rightarrow$ n (udd) + π^+ (ud`)** is **0.4831 or 48.31 %.**

That is, for each 10000 Σ^+ (uus) produced, 5157 will decay into p (uud) + π^0 (uu`), and 4831 into n (udd) + $\pi^+$ (ud`).

The **branching ratio** for the decay of tauon τ^- into an electron e^-, an electron-type antineutrino v_e` and a tauon-type neutrino v: **$\tau^- \rightarrow e^- + v_e$` + v_τ** is **0.1783 or 17.83 %.**

The **branching ratio** for the decay: **$\tau^- \rightarrow \mu^- + v_\mu$` + v_τ** is **0.1741 or 17.41 %.**

The **branching ratio** for the decay: **$\tau^- \rightarrow \pi^-$ (u`d) + v_τ** is **0.6482 or 64.82 %.**

That is, for each 100 tauon τ^- produced, 17 will decay into e^- + v_e` + $v_\tau$, 17 into $\mu^-$ + $v_\mu$` + v_τ, and 64 into π^- (u`d) + v_τ.

The **branching ratio** for the decay of the negative sigma baryon **Σ^- (dds) into a neutron n (udd) and a negative pion π^- (u`d): $\Sigma^-$ (dds) $\rightarrow$ n (udd) + $\pi^-$ (u`d)** is **0.99848 or 99.848 %.**

The **branching ratio** for the decay: Σ^- **(dds)** $\rightarrow$ **n (udd) + e$^-$ + v$_e$`** is 1.017×10^{-3} **or** 0.1017 %.

That is, for each 100000 Σ^- **(dds)** produced, 99848 will decay into **n (udd) + π^- (u`d)** and 101 into **n (udd) + e$^-$ + v$_e$`**

The **branching ratio** for the decay of the positive pion π^+ (ud`) into an antimuon $\mu^+$ and a muon-type neutrino $v_\mu$: $\pi^+$ **(ud`)** $\rightarrow$ **μ^+ + v$_\mu$** is 0.999877 or 99.9877 %.

The **branching ratio** for the decay of the positive pion π^+ (ud`) into a positron and an electron-type neutrino $v_e$: $\pi^+$ **(ud`)** $\rightarrow$ **e$^+$ + v$_e$** is 1.23×10^{-4} or 0.0123 %.

That is, for each 1000000 π^+ (ud`) produced, 123 will decay into e$^+$ + v$_e$, and all others into μ^+ + v$_\mu$.

The **branching ratio** for the decay of the muon μ^- into an electron e$^-$, an electron-type antineutrino v_e` and a muon-type neutrino $v_\mu$: $\mu^-$ $\rightarrow$ **e$^-$ + v$_e$` + v$_\mu$** is 1 or 100 %.

That is, each muon μ^- always decay into e$^-$ + v$_e$` + v$_\mu$.

note: The partial width of a given decay channel is the product of the total width and the corresponding branching ratio. For example, the total width of the W boson is 2.08 GeV and the branching ratio for the decay: $W^+ \rightarrow e^+ + v_e$ is 0.107 or 10.7 %. Thus, **the partial width Γ_{ev} of the process:** $W^+ \rightarrow e^+ + v_e$ **is 0.107 x 2.08 GeV = 0.222 GeV = 222 MeV.** The sum of the partial widths of a particle equals the total width. For example, in case of W^+ boson, $\Gamma_{TOL} = \Gamma_{ev} + \Gamma_{\mu v} + \Gamma_{\tau v} + \Gamma_{hadrons}$.

Note that the branching ratio for the process: $W^+ \rightarrow e^+ + v_e = \Gamma_{ev} / \Gamma_{TOL}$

Cross-Section and Resonance

The cross-section is more often used to mean the **probability** that two particles will collide and react in a certain way. For instance, when physicists measure the "proton-proton to top-antitop" cross-section, they are counting how many top-antitop pairs were produced, when a given number of protons were collided.

The cross-section is the **ratio** of the number of new particles/states produced and the number of particles collided.

Suppose, we do the different experiments with the **same number of colliding particles** in each experiment but in each successive experiment, the cms energy of colliding particles is greater than that in the last experiment. Then the cross-section is proportional to the number of new particles/states produced.

To understand the cross-section and the resonance, consider the following example:

An electron-positron pair e^+e^- annihilates to a Z^0 boson which then decays into a muon-antimuon pair $\mu^+\mu^-$.

note: The branching ratio for the decay: $Z^0 \rightarrow \mu^+ + \mu^-$ is about .0336 or 3.36%, i.e. out of every 10000 Z^0 bosons produced, 336 will decay into muon-antimuon pairs $\mu^+\mu^-$.

Consider an electron-positron collider. We can control the centre-of-mass energy E_{cms} of the electron-positron collision.

Suppose, the different experiments are done with the **same number of electrons and positrons** in each experiment but in each successive experiment, the cms energy of colliding electron-positron pairs is greater than that in the last experiment.

In each experiment, a beam of electrons consisting of 10 bunches (successive bunches being separated by say, 5 meters), each bunch having 1 billion accelerated electrons may be collided with a similar beam of positrons coming from the opposite direction.

The first bunch of the electrons will collide with the first bunch of the positrons and after the collision, they will scatter in different directions while producing, for example 0 or 1 or 5 or 10 or 40 or 50 or 100 or more real Z^0 bosons depending upon the centre-of-mass energy E_{cms} of the colliding electron-positron pairs.

The second bunch of the electrons will collide with the second bunch of the positrons. The third bunch of the electrons will collide with the third bunch of the positrons and similarly, the tenth bunch of the electrons will collide with the tenth bunch of the positrons.

If the centre-of-mass energy E_{cms} of the colliding electron-positron pair is such that the collision of a bunch of electrons and a bunch of positrons produced 3 Z^0 bosons, then in the entire experiment involving 10 bunches of electrons and 10 bunches of positrons, on average, 30 Z^0 bosons will be produced.

note: After the collision, via detectors, we can count the number of final states produced (in this case, $\mu^+\mu^-$ pair).

note: If in an experiment, during the collisions of bunches of electrons and bunches of positrons, 336 muon-antimuon pairs are produced, then it means collisions of those bunches of electrons and positrons caused 10000 electron-positron pairs to produce 10000 Z^0 bosons, the energy of each Z^0 boson being equal to the cms energy of an electron-positron

pair that is, the entire cms energy of each of those 10000 electron-positron pairs converted into the total energy of a single Z^0 boson.
note: Collision of electron-positron pair does not mean one electron-positron pair is colliding at a particular time, it means, collision of a bunch of electrons and a bunch of positrons so that millions or billions of electron-positron pairs are colliding at the same time.

Now, suppose, in our **first experiment, the colliding electron-positron pairs** are accelerated to such an extent that the cms energy **of colliding electron-positron pairs is ~ 88.2 GeV** that is, each electron and each positron has energy ~ 44.1 GeV before the collision. Suppose, in this first experiment involving a beam of the electrons and a similar beam of the positrons, $m_{88.2}$ = 570 Z^0 bosons are produced (each with energy 88.2 GeV), out of which, 3.36%, i.e. 19 Z^0 bosons decay to produce $n_{88.2}$ = 19 muon-antimuon pairs $\mu^+\mu^-$.

Now, suppose, we do the **second experiment** with the similar beams of the electrons and the positrons, but **the colliding electron-positron pairs** are accelerated to such an extent that the cms energy **of colliding electron-positron pairs is ~ 88.4 GeV** that is, each electron and each positron has energy ~ 44.2 GeV before the collision. Suppose, in this second experiment, $m_{88.4}$ = 715 Z^0 bosons are produced (each with energy 88.4 GeV), out of which, 3.36%, i.e. 24 Z^0 bosons decay to produce $n_{88.4}$ = 24 muon-antimuon pairs $\mu^+\mu^-$.

Similarly, in the third, fourth, fifth, sixth …….. 12th experiments, suppose the colliding electron-positron pairs are accelerated to such an extent that the **cms energy of colliding electron-positron pairs is** about 89.2 GeV, 89.5 GeV, 90.2 GeV, 91.2 GeV, 92 GeV, 92.2 GeV, 92.8 GeV, 93 GeV, 93.4 GeV, 94 GeV respectively.

Suppose, in these experiments, $m_{89.2}$ = 1300 (each with energy 89.2 GeV), $m_{89.5}$ = 1430 (each with energy 89.5 GeV), $m_{90.2}$ = 2700 (each with energy 90.2 GeV), $m_{91.2}$ = 4300 (each with energy 91.2 GeV), m_{92} = 3400 (each with energy 92 GeV), $m_{92.2}$= 3000 (each with energy 92.2 GeV), $m_{92.8}$ = 2100 (each with energy 92.8 GeV), m_{93} = 2000 (each with energy 93 GeV), $m_{93.4}$ = 1400 (each with energy 93.4 GeV), and m_{94} = 1300 (each with energy 94 GeV) Z^0 bosons are produced respectively, and as 3.36% of the total Z^0 bosons produced decay into muon-antimuon pairs $\mu^+\mu^-$, thus, in third, fourth, fifth, sixth …….. 12th experiment, $n_{89.2}$ = 43, $n_{89.5}$ = 48, $n_{90.2}$ = 90, $n_{91.2}$ = 144, n_{92} = 114, $n_{92.2}$ = 100, $n_{92.8}$ = 70, n_{93} = 67, $n_{93.4}$ = 47, n_{94} = 43 muon-antimuon pairs $\mu^+\mu^-$ will be produced (through the decay of the Z^0 bosons) respectively. These results imply:

$m_{91.2}$ **(4300)** > $m_{90.2}$ (2700) > $m_{89.5}$ (1430) > $m_{89.2}$ (1300) > $m_{88.4}$ (715) > $m_{88.2}$ (570)

$m_{91.2}$ **(4300)** > m_{92} (3400) > $m_{92.2}$ (3000) > $m_{92.8}$ (2100) > m_{93} (2000) > $m_{93.4}$ (1400) > m_{94} (1300)

and

$n_{91.2}$ **(144)** > $n_{90.2}$ (90) > $n_{89.5}$ (48) > $n_{89.2}$ (43) > $n_{88.4}$ (24) > $n_{88.2}$ (19)

$n_{91.2}$ **(144)** > n_{92} (114) > $n_{92.2}$ (100) > $n_{92.8}$ (70) > n_{93} (67) > $n_{93.4}$ (47) > n_{94} (43)

$m_{88.4}$ **(715)** > $m_{88.2}$ **(570)** and $n_{88.4}$ **(24)** > $n_{88.2}$ **(19) imply** that the number $m_{88.4}$ (715) of the Z^0 bosons and therefore the number $n_{88.4}$ (24) of $\mu^+\mu^-$ pairs produced at electron-positron cms energy of 88.4 GeV are **more** than the number $m_{88.2}$ (570) of the Z^0 bosons and therefore the number $n_{88.2}$ (19) of $\mu^+\mu^-$ pairs produced at electron-positron cms energy of 88.2 GeV.

Similarly, $m_{89.2}$ **(1300)** > $m_{88.4}$ **(715)** and $n_{89.2}$ **(43)** > $n_{88.4}$ **(24) imply** that the number $m_{89.2}$ (1300) of the Z^0 bosons and therefore the number $n_{89.2}$ (43) of $\mu^+\mu^-$ pairs produced at electron-positron cms energy of 89.2 GeV are **more** than the number $m_{88.4}$ (715) of the Z^0 bosons and therefore the number $n_{88.4}$ (24) of $\mu^+\mu^-$ pairs produced at electron-positron cms energy of 88.4 GeV.

Similarly, $m_{89.5}$ **(1430)** > $m_{89.2}$ **(1300)** and $n_{89.5}$ **(48)** > $n_{89.2}$ **(43)** imply that the number $m_{89.5}$ (1430) of the Z^0 bosons and therefore the number $n_{89.5}$ (48) of $\mu^+\mu^-$ pairs produced at electron-positron cms energy of 89.5 GeV are **more** than the number $m_{89.2}$ (1300) of the Z^0 bosons and therefore the number $n_{89.2}$ (43) of $\mu^+\mu^-$ pairs produced at electron-positron cms energy of 89.2 GeV.

The same pattern for the cms energy of 90.2 GeV, 91.2 GeV.

However, $m_{91.2}$ **(4300)** > m_{92} **(3400)** and $n_{91.2}$ **(144)** > n_{92} **(114)** imply that the number m_{92} (3400) of the Z^0 bosons and therefore the number n_{92} (114) of $\mu^+\mu^-$ pairs produced at electron-positron cms energy of 92 GeV are **less** than the number $m_{91.2}$ (4300) of the Z^0 bosons and therefore the number $n_{91.2}$ (144) of $\mu^+\mu^-$ pairs produced at electron-positron cms energy of 91.2 GeV.

Similarly, m_{92} **(3400)** > $m_{92.2}$ **(3000)** and n_{92} **(114)** > $n_{92.2}$ **(100)** imply that the number $m_{92.2}$ (3000) of the Z^0 bosons and therefore the number $n_{92.2}$ (100) of $\mu^+\mu^-$ pairs produced at electron-positron cms energy of 92.2 GeV are **less** than the number m_{92} (3400) of the Z^0 bosons and therefore the number n_{92} (114) of $\mu^+\mu^-$ pairs produced at electron-positron cms energy of 92 GeV.

Similarly, $m_{92.2}$ **(3000)** > $m_{92.8}$ **(2100)** and $n_{92.2}$ **(100)** > $n_{92.8}$ **(70)** imply that the number $m_{92.8}$ (2100) of the Z^0 bosons and therefore the number $n_{92.8}$ (70) of $\mu^+\mu^-$ pairs produced at electron-positron cms energy of 92.8 GeV are **less** than the number $m_{92.2}$ (3000) of the Z^0 bosons and therefore the number $n_{92.2}$ (100) of $\mu^+\mu^-$ pairs produced at electron-positron cms energy of 92.2 GeV.

The same pattern for the cms energy of 93 GeV, 93.4 GeV and 94 GeV.

The number $m_{91.2}$ (4300) of the Z^0 bosons and therefore the number $n_{91.2}$ (144) of $\mu^+\mu^-$ pairs produced are maximum at electron-positron cms energy of 91.2 GeV.

As the cms energy of the colliding electron-positron pair is increased from 88.2 GeV to, say 89.2 GeV, there is an **increase in the production of final states** (in this case, $\mu^+\mu^-$ pairs), and **even more** $\mu^+\mu^-$ pairs are produced at electron-positron cms energy of 90.2 GeV and the number of $\mu^+\mu^-$ pairs produced reaches to the **maximum** at electron-positron cms energy equal to the rest mass (91.2 GeV) of the Z^0 boson and the number of $\mu^+\mu^-$ pairs produced begin to **decrease** as we do further experiments with electron-positron cms energy greater than 91.2 GeV.

This means the **cross-section increases as the cms energy of colliding electron-positron pair increases from 88.2 GeV to 91.2 GeV and then decreases as the cms energy of colliding electron-positron pair increases beyond 91.2 GeV.**

Thus, **at the cms energy of 91.2 GeV, there is a peak, i.e. resonance** in the graph drawn between 'the cross-section (along Y axis)' versus 'the cms energy of colliding particles (along X-axis)' which **implies there exists in nature a massive particle with rest mass of 91.2 GeV.**

Thus, the position of the resonance tells us the mass of that particle which in this case is the Z^0 boson. Thus, the idea of the resonance gives a simple way to detect the Z^0 boson even if it decays before it reached the detectors.

The simplest explanation of the resonance particles, or resonances is that they are extremely short lived particles. The mean lifetime of these particles is of the order of 10^{-23} seconds. Travelling at the speed of light, these particles could only travel about 10^{-15} meters before decaying.

In fact, one way to search for the new particles is to follow the above method that is, scan over energies, and see, if a peak is observed in the graph. The location of the resonance or peak tells us the mass of the intermediate particle. Mass may also be deduced through detectors and counters which analyze the final products (in which resonance particle decays).

In the collisions of the electron-positron pairs at various energies ranging from 100 MeV to 10 GeV, there exist several peaks or resonances in the graph between the cross-section (along Y-axis) versus the cms energy of colliding electron-positron pairs (along X-axis). **These peaks correspond to the various vectors bosons: ρ^0 (770 MeV), ω (782 MeV), φ (1020 MeV), J/ψ (3.1 GeV), Υ (9.46 GeV).**

(note that even mesons are bosons).
The collisions of the electron-positron pairs led to the discovery of these vector mesons (except upsilon meson) and many other massive particles. For example, when the cms energy of colliding electron-positron pairs was 3.1 GeV, a peak was found in the graph which implied there exists in nature a particle with rest mass 3.1 GeV. This particle was called the J/ψ (cc`).

<u>The cross-section decreases inversely to the cms energy of colliding e$^+$e$^-$ pairs</u> **provided** the number of colliding e$^+$e$^-$ pairs are fixed (e.g. 10 billions) in each of those experiments.

Thus, whereas the cross-section is very large at low cms energy, say 1 GeV, it is extremely small at the cms energy, say 90 GeV.

The lesser the cross-section for a resonance particle, the lesser the number of that particles produced during the collisions of the e^+e^- pairs. For example, the **cross-section** for the 'resonance particle **rho ρ^0 meson** (rest mass: +770 MeV) at the electron-positron cms energy 770 MeV' is **somewhat larger than** the cross-section for the 'resonance particle **phi φ meson** (rest mass: 1020 MeV) at the electron-positron cms energy 1020 MeV'. **This also means, the numbers of** ρ^0 mesons produced, if 10 billion electron-positron pairs were collided at the cms energy 770 MeV (the rest mass of ρ^0), are **somewhat greater** than the numbers of φ mesons produced, if 10 billion electron-positron pairs were collided at the cms energy 1020 MeV (the rest mass of φ).

However, the cross-section for the 'resonance particle **rho ρ^0 meson** (rest mass: +770 MeV) at the electron-positron cms energy 770 MeV' is **much larger than** the cross-section for the 'resonance particle **upsilon Υ meson** (rest mass: 9.46 GeV) at the electron-positron cms energy 9.46 GeV. **This also means, the numbers of** ρ^0 mesons produced, if 10 billion electron-positron pairs were collided at the cms energy 770 MeV (the rest mass of ρ^0), are **much greater** than the numbers of Υ mesons produced, if 10 billion electron-positron pairs were collided at the cms energy 9.46 GeV (the rest mass of Υ)'.

Similarly, the cross-section for the 'resonance particle **rho ρ^0 meson** (rest mass: +770 MeV) at the electron-positron cms energy 770 MeV' is **extremely larger than** the cross-section for the 'resonance particle Z^0 boson (rest mass: 91.2 GeV) at the electron-positron cms energy 91.2 GeV. **This also means, the numbers of** ρ^0 mesons produced, if 10 billion electron-positron pairs were collided at the cms energy 770 MeV (the rest mass of ρ^0), are **extremely greater** than the numbers of Z^0

bosons produced, if 10 billion electron-positron pairs were collided at the cms energy 91.2 GeV (the rest mass of Z^0)'.

Whereas only 1 top–antitop quark pair is produced in the collisions of 10 billion proton-proton pairs at the cms energy 1800 GeV, several Z^0 bosons would be produced, if 10 billion electron-positron pairs were collided at the cms energy 91.2 GeV. Similarly, the total number of Υ (9.46 GeV) produced will be even larger if 10 billion electron-positron pairs were collided at the cms energy 9.46 GeV. The same pattern for J/ψ (3.1 GeV), φ (1020 MeV), ω (782 MeV) and ρ^0 (770 MeV) with 'the number of ρ^0 mesons (lightest out of all these mesons) produced to be largest, if 10 billion electron-positron pairs were collided at the cms energy 770 MeV (the rest mass of ρ^0)'.

Suppose,

m_1 ρ^0 mesons are produced, if 10 billion electron-positron pairs were collided at the cms energy 770 MeV (the rest mass of ρ^0)'.

m_2 ω mesons are produced, if 10 billion electron-positron pairs were collided at the cms energy 782 MeV (the rest mass of ω)'.

m_3 φ mesons are produced, if 10 billion electron-positron pairs were collided at the cms energy 1020 MeV (the rest mass of φ).

m_4 J/ψ mesons are produced, if 10 billion electron-positron pairs were collided at the cms energy 3.1 GeV (the rest mass of J/ψ),

m_5 Υ mesons are produced, if 10 billion electron-positron pairs were collided at the cms energy 9.46 GeV (the rest mass of Υ).

m_6 Z^0 bosons are produced, if 10 billion electron-positron pairs were collided at the cms energy 91.2 GeV (the rest mass of Z^0).

Then $m_1 > m_2 > m_3 > m_4 > m_5 > m_6$.

The peaks are the evidences for the presence of the resonance particles which form as intermediate steps in the collisions. **Different peaks are caused by a large number of resonance particles which are as real as other particles, the only difference being a difference in lifetime. Resonance particles have extremely small lifetime and therefore are highly unstable.**

A resonance particle does not have a unique mass but a distribution with width $\Gamma = \hbar/\tau$.

When mean lifetime τ is very short, width Γ is large. For example, Z^0 boson has central mass 91.2 GeV (which we usually say as rest mass) and has a larger width $\Gamma = 2.49$ GeV.

Thus, when an electron-positron pair having cms energy, say 89 GeV annihilates to a Z^0 boson, then that Z^0 boson will have mass 89 GeV which is less than the rest mass (91.2 GeV) of the real Z^0 boson. Similarly, when an electron-positron pair having cms energy, say 93 GeV annihilates to a Z^0 boson, then that Z^0 boson will have mass 93 GeV which is larger than the rest mass (91.2 GeV) of the real Z^0 boson. If the cms energy of the colliding e^+e^- pairs is 91.2 GeV, then the number of Z^0 bosons produced would be maximum. However, if the cms energy E_{cms} of colliding e^+e^- pairs is **91.2 GeV - $\Gamma/2$** = 91.2 - 1.245 **= 89.95 GeV**, then the number of Z^0 bosons produced would be **half its maximum value.** (If 4300 Z^0 bosons are produced at the cms energy of 91.18 GeV, then 4300/2 = 2150 Z^0 bosons will be produced at the cms energy of 91.18 - $\Gamma/2$ = 89.94 GeV). The resonance curve is not exactly symmetrical on the right and the left side of the peak value of the numbers of Z^0 bosons produced. Thus, if the cms energy of colliding e^+e^- pairs is **91.2 GeV + $\Gamma/2$** = 91.2 + 1.245 GeV **= 92.44 GeV**, then the number of Z^0 bosons produced will be somewhat more or less than half its maximum value.

Thus, the **probability** of the production of Z^0 bosons is large, if the cms energy of e^+e^- pairs is in the range of **90 GeV to 92.5 GeV** and relatively less for E_{cms} < 90 GeV and E_{cms} > 92.5 GeV and at the cms energy of e^+e^- pairs less than say, 75 GeV and greater than, say 100 GeV, no Z^0 boson is produced. That is, **the mass of Z^0 boson varies from ~ 75 GeV to ~ 100 GeV.**

The cross-section for a reaction (in this case, the production of the Z^0 boson) is proportional to the term $1/\{(E - M_z)^2 - \Gamma^2\}$, where M_z (91.2 GeV) is the rest mass of the Z^0 boson, E is the centre-of-mass energy of the colliding electron-positron pair, and Γ (2.49 GeV) is the decay width of the Z^0 boson. When electron-positron pairs have the cms energy E **equal to** the rest energy M_z of the Z^0 boson, $E - M_z = 0$ and so the cross-section looks like $1/\Gamma$.

Thus, for a small width Γ (small in comparison with the central mass of resonance particle), the cross–section for this process becomes very large that is, the number of the Z^0 bosons produced, 'when the cms energy of colliding electron-positron pairs **is equal** to the the rest energy of the Z^0 boson' are **much more** than the number of the Z^0 bosons produced, 'when the cms energy of colliding electron-positron pairs is **less or more** than the rest energy of the Z^0 boson'.

note:

The Z^0 bosons produced during the collisions of e^+e^- pairs at and around the cms energy of 91 GeV are real, just as vector mesons e.g. ρ^0, ϕ, Υ produced are real.

The real W bosons (which are charged particles) cannot be produced through the collisions of electron-positron pairs, as net electric charge of e^+e^- pair is zero. That is, during the collisions of e^+e^- pairs at and around the cms energy of 80 GeV (the rest mass of W boson), there does not exist any resonance state. In fact, in the collisions of e^+e^-

pairs, the next resonance state after Υ (9.46 GeV) is Z^0 (91.2 GeV).

note:

W^+, W^-, Z^0 bosons produced during the collisions of pp` pairs are also real.

The real W bosons are also produced by the decay of the top-antitop quark pairs. Higgs bosons may also decay into real W and Z bosons.

W^+W^- and Z^0 pair-production

The real W and Z bosons were also produced in **pairs** during the collision of e^+e^- pairs.

As the mass of the W boson is about 80 GeV, the threshold energy for the W^+W^- pair-production is about 161 GeV (little more than the sum of the masses of the W^+W^- bosons).

In the e^+e^- collisions, the **W^+W^- pairs (the real W bosons) were first produced** in the LEP (Large Electron-Positron) circular Collider at CERN in 1997 via neutrinos, photons and Z^0 exchanges, when e^+e^- pairs were collided at the cms energy of 161 GeV and more.

Although number of W^+W^- pairs produced increased as experiments were done by increasing the cms energy of colliding e^+e^- pairs step by step, however, almost the same number of W^+W^- pairs was produced at the electron-positron cms energy in the range 172 GeV - 209 GeV. (Maximum cms energy obtained in the LEP was 209 GeV). Thus, **W^+W^- pair-production is not a resonance state.**

DELPHI detector at the LEP collider detected W^+W^- pairs.

In many experiments, one W boson underwent **hadronic** decay, e.g. $W^+ \rightarrow \pi^+$ (ud`) or $W^+ \rightarrow K^+$ (us`), and other underwent **leptonic** decay e.g $W^- \rightarrow \mu^- + v_\mu$`.

As the mass of the Z boson is about 91 GeV, the threshold energy for the Z^0 pair-production is about 182 GeV.

In the e^+e^- collisions, the **Z^0 pairs (the real Z bosons) were first produced** in the LEP (Large Electron-Positron) circular Collider at CERN when e^+e^- pairs were collided at the cms energy of 182 GeV and more.

In many experiments, **one Z^0 boson decayed into <u>bottom-antibottom pair</u> bb` which then decayed weakly into B mesons e.g. into a positive B meson and a negative pion:. bb` $\rightarrow$ B$^+$ (ub`) + $\pi^-$ (u`d) which** is a hadronic decay:

bb` $\rightarrow$ B$^+$ (ub`) + π^- (u`d).

Here, the **bottom quark** (charge: -1/3e) of the bottom-antibottom pair bb` **emits a W$^-$ boson and transforms into** an **up quark u** (charge: +2/3e) and the bottom-antibottom pair transforms into a **positive B meson B$^+$ (ub`).**

The **W$^-$ boson subsequently decays into an antiup quark u` (charge: -2/3e) and a down quark d (charge: -1/3e) which** combine to produce a **negative pion $\pi^-$ (u`d).**

Another Z^0 boson underwent leptonic decay e.g $Z^0 \rightarrow \mu^+ + \mu^-$

Discovery of the Vector Mesons

Examples of Vector Mesons: ρ^0 - $(u\bar{u}$- $d\bar{d})/2^{\frac{1}{2}}$ (rest mass: 770 MeV), ω - $(u\bar{u}$+ $d\bar{d})/2^{\frac{1}{2}}$ (rest mass: 782 MeV), ϕ – $s\bar{s}$ (rest mass: 1020 MeV), **J/ψ** – $c\bar{c}$ (rest mass: 3.1 GeV), Υ – $b\bar{b}$ (rest mass: 9.46 GeV)

In the collisions of the electron-positron pairs at various energies ranging from 100 MeV to 10 GeV, there exist several peaks or resonances in the graph between the cross-section (along Y-axis) versus the cms energy of colliding electron-positron pairs (along X-axis). **These peaks correspond to the various vectors bosons: ρ^0 (770 MeV), ω (782 MeV), φ (1020 MeV), J/ψ (3.1 GeV), Υ (9.46 GeV).**

(note that even mesons are bosons).

note: There are peaks at 3.7 GeV, 3.77 GeV, 10.02 GeV, 10.36 GeV, 10.58 GeV too corresponding to ψ(2S), ψ(3S), Υ(2S), Υ(3S), Υ(4S) respectively.

The most elegant method for the production of the vector mesons is the colliding electron and positron beams.

At the electron-positron pair cms energy E_{cms} < 10 GeV, **the annihilation of e^+e^- into the vector mesons (ρ^0, ω, φ, J/ψ, Υ)** proceeds dominantly through a single virtual photon. The ability of high-energy virtual photons to take a broad range of masses allow them to take on the masses and attributes of certain hadrons, called vector mesons.

The spin (J =1) and parity (odd) are the same for the photons and the vector mesons. Thus, the virtual photon can easily convert into a vector meson provided it has the energy (obtained through the annihilation of e⁺e⁻ pair) equal to or about the rest mass of that vector meson. The vector mesons were among the first of nearly one hundred resonance states of baryons and mesons discovered.

When an electron collides with a positron, the mass and energy of these beam particles may convert into the energy of a single 'massive' virtual photon. This virtual photon may then convert in a very short time to a vector meson (quark-antiquark pair).

To begin with, e⁺e⁻ come from the opposite directions (180^0 with respect to each other), collide, and annihilate to a virtual photon. This virtual photon subsequently decays into quark-antiquark pair QQ` i.e. into a vector meson (ρ^0, ω, φ, J/ψ or Υ).

As the virtual photon produced in the head-on collision of e⁺e⁻ has no momentum, the quark and the antiquark (uu`, dd`, ss`, cc` or bb`) of the vector meson (ρ^0, ω, φ, J/ψ or Υ) produced by the decay of a virtual photon also move in the opposite directions (180^0 with respect to each other) with identical momentum but at an angle with respect to e⁺e⁻ axis.

Each of the quark and the antiquark of the vector meson, after some time decays into the hadrons and therefore the hadrons appear in the form of two oppositely directed jets collimated around the QQ` axis.

Consider the production of the rho meson **ρ^0**.

If the colliding electron-positron pairs are accelerated to such an extent that the cms energy **of colliding electron-positron pairs is** equal to or about 770 MeV before the collision, then the single virtual photon produced through the annihilation of an e⁺e⁻ pair possesses the entire energy of that e⁺e⁻ pair, i.e. the cms energy (~770 MeV) of e⁺e⁻ pair becomes the total energy or the mass of the virtual photon. This **virtual photon (having mass ~770 MeV) subsequently decays into an up and an antiup quark pair uu` or into a down and an antidown quark pair dd`, i.e. into rho meson ρ⁰** (rest mass: 770 MeV) which **quickly decays into the pions.**

Suppose, the different experiments are done with the **same number of electrons and positrons** in each experiment but in each successive experiment, the cms energy of colliding electron-positron pairs is greater than that in the last experiment.

Now, suppose, in our **first experiment, the colliding electron-positron pairs** are accelerated to such an extent that the cms energy **of colliding electron-positron pairs is 660 MeV** that is, each electron and each positron has energy 330 MeV before the collision.
Similarly, in the second, third, fourth, fifth, sixth experiments, suppose the colliding electron-positron pairs are accelerated to such an extent that the **cms energy of colliding electron-positron pairs is** 700 MeV, 770 MeV, 800 MeV, 820 MeV, 880 MeV respectively.
Suppose, through the detectors and counters, it is found that the number of pion pairs produced in the first, second, third, fourth, fifth, sixth experiments are around 12, 25, 55, 45, 20, 10 respectively.

Thus, the number of the pion pairs (55) produced are maximum in the third experiment , i.e. at the electron-positron cms energy of 770 MeV.

Thus, **at the cms energy of 770 MeV, there is a peak, i.e. resonance** in the graph drawn between 'the cross-section (along Y axis)' versus 'the cms energy of colliding electron positron pairs (along X-axis)' which **implies there exists in nature a massive particle with rest mass of 770 MeV** (which decayed into pion pairs). This particle was called the rho meson.

The cross-section is equivalent to the number of final particles produced after collisions which in this case are pions.

Similarly, there are peaks at E_{cms} of 782 MeV, 1020 MeV, 3.1 GeV, 9.46 GeV corresponding to **ω, φ, J/ψ, ϒ respectively.**

For example, **if the colliding electron-positron pairs** are accelerated to such an extent that the cms energy **of colliding electron-positron pairs is** equal to or about 1020 MeV before the collision, then an e^+e^- pair may annihilate to a virtual photon and the entire cms energy (~1020 MeV) of that e^+e^- pair becomes the total energy or the mass of the virtual photon. This **virtual photon (having mass ~1020 MeV) subsequently decays into a strange and an antistrange quark pair ss`, i.e. into phi meson φ** (rest mass: 1020 MeV) which quickly decays into the kaons.

The strange quark and the antistrange quark of this phi meson **φ (ss`)** move in the opposite directions (180^0 with respect to each other) with identical momentum but at an angle with respect to e^+e^- axis.

The annihilation of e^+e^- into the vector mesons (ρ^0, ω, φ, J/ψ, ϒ) proceeds dominantly through a single virtual photon.

Quark and antiquark pair of such a vector meson e.g. strange quark s and antistrange quark s`pair of phi meson φ are unstable and quickly decay into observable particles called hadrons (baryons or mesons). This process is called hadronization.

Now, consider the following decay. This is 29th example in the topic: Examples of the Decays.

29. The decay of the phi meson φ (ss`) into a positive kaon K⁺ (us`) and a negative kaon K⁻ (u`s):

φ (ss`) → K⁺ (us`) + K⁻ (u`s)

As the separation between the strange quark s and the antistrange quark s` (moving in the opposite directions) of the phi meson φ (ss`) increases, the colour or strong force between them gets stronger and stronger and after being separated by say 2 fermi (1 fermi = 10^{-15} meters), a gluon is emitted by the strange quark or the antistrange quark. This gluon decays into the up quark and the antiup quark. However, the **up quark u and the antiup quark u` produced by the decay of the gluon cannot combine to produce the neutral pion π⁰ (uu`)** because these up and antiup quarks too move in the opposite directions. The **strange quark s** of the phi meson and the **antiup quark u`** produced by the decay of the gluon move in the same direction and combine to produce a **negative kaon K⁻ (u`s).** The **antistrange quark s`** of the phi meson and the **up quark u** produced by the decay of the gluon move in the same direction with respect to each other but at 180⁰ with respect to those quark-antiquark which combine to produce negative kaon K⁻ (u`s). These **antistrange quark s` and up quark u** combine to produce **positive kaon K⁺ (us`).** This means, **the positive kaon K⁺ (us`) moves at 180⁰ with respect to the negative kaon K⁻ (u`s).**

In this way, the **original strange quark and antistrange quark pair,** i.e. phi meson φ (ss`) ends up as two kaons (which are mesons). The kaons emerge in two back-to-back jets, one along the direction of the primordial strange quark, the other along the direction of the primordial antistrange quark.

That is, the **original quark and antiquark pair**, i.e. a vector meson (ρ^0, **ω, φ, J/ψ or ϒ**) evebtually ends up as hadrons (baryons and/or

mesons). The hadrons emerge in two back-to-back jets, one along the direction of the primordial quark, the other along the direction of the primordial antiquark.

By detecting these hadrons in detectors, information is obtained about the quark-antiquark pairs which decayed into those hadrons.

Sometimes, a quark may radiate a 'hard' gluon (high energy gluon), carrying about half of the quark energy, at a large angle. This quark and the gluon emitted by this quark give rise to a distinct third hadronic jet in the detector, separate from the jets formed by the original quark and antiquark pair. Thus, the three jet events provided the evidence of the existence of gluon inside nucleon. The angular distribution in three jet events is sensitive to the gluon spin, and the analysis of three jet events predicted that the gluon has spin parity $J^P =$ 1^-. **The observation of the three-jet events is generally regarded as the most direct evidence for the existence of the gluons.**

Three-jet events were first observed in TASSO (Two Arm Spectrometer Solenoid) detector at PETRA (Positron-Electron Tandem Ring Accelerator) which resulted in the discovery of the gluon in 1979.

PETRA was a powerful electron-positron collider at the German national laboratory DESY (Deutsches Elektronen-Synchrotron) in Hamburg, Germany. It provided the high-energy collisions necessary for TASSO and other experiments to probe fundamental physics and study the behavior of particles at high energies.

Three-jet events have been studied extensively at various colliders, including the LEP collider.

An electron-positron pair annihilates to a virtual photon which subsequently decays into, i.e. produces quark-antiquark pair: $e^+e^- \rightarrow \gamma \rightarrow Q\bar{Q}$. This quark-antiquark pair is a vector meson (e.g. ρ, ω, φ or **J/ψ**)

The quark-antiquark pair $Q\bar{Q}$ i.e. vector meson (e.g. ρ, ω, φ or **J/ψ**) then decays to hadrons: $Q\bar{Q} \rightarrow$ Hadrons.

Experiments found that the annihilation of e^+e^- to hadrons proceed as a point like process as does $e^+e^- \rightarrow \mu^+\mu^-$. The pointlike constituents or partons are called the quarks.

Collisions of e^+e^- pairs may produce only those vector mesons which are neutral, as net electric charge of e^+e^- pair is zero. However, **these neutral** <u>**vector**</u> **mesons (e.g. ρ^0, ω, φ) decay into charged and neutral** <u>**pseudoscalar**</u> **mesons (kaons, pions).**

Typically, the neutral rho meson decays into charged pions ($\rho^0 \rightarrow \pi^+ + \pi^-$), the omega meson decays into charged and neutral pions ($\omega \rightarrow \pi^+ + \pi^- + \pi^0$) and the phi meson decays into charged and neutral kaons ($\varphi \rightarrow K^+ + K^-$ or $K^0 + \bar{K^0}$).

All these decays are strong decays, i.e. through gluons.

Rho meson ρ^0 (mean lifetime $\tau = 4.4 \times 10^{-24}$ s) decays into π^+ ($u\bar{d}$) + π^- ($\bar{u}d$)

Omega meson ω ($\tau = 7.7 \times 10^{-23}$ s) decays into π^+ ($u\bar{d}$) + π^- ($\bar{u}d$) + π^0 ($u\bar{u}$) or π^0 ($u\bar{u}$) + γ

Phi meson φ ($\tau = 1.5 \times 10^{-22}$ s) decays into K^+ ($u\bar{s}$) + K^- ($\bar{u}s$) or K^0 ($d\bar{s}$) + $\bar{K^0}$ ($\bar{d}s$)

Mean lifetimes of **ρ⁰, ω, φ** (10^{-24} s, 10^{-23} s, 10^{-22} s respectively) is so small (though much longer than the lifetime of a virtual photon) that they decay before travelling (with velocity nearly equal to that of light) enough distance to leave their traces in detectors. For example, these vector mesons cannot form a track in the bubble chamber, hence they are observed indirectly through the long-lived particles into which they decay. Their decay products, i.e. **charged pions π⁺ (ud`), π⁻ (u`d)** having mean lifetime $\tau = 2.6 \times 10^{-8}$ s and **charged kaons K⁺ (us`), K⁻ (u`s)** having mean lifetime $\tau = 1.2 \times 10^{-8}$ s can travel some distance with velocity nearly equal to that of light and **leave their traces in detectors.**

In that topic, Discovery of Strange Quark, there are following lines:
Mean lifetime of lambda baryon Λ⁰ (uds) is $\tau \sim 2.6 \times 10^{-10}$ s, and $\tau \sim 2.6 \times 10^{-10}$ s after the interaction of **π⁻ (u`d)** and **p (uud)**, somewhere else, a pair of track appeared, when **Λ⁰ (uds) decayed** into a proton p (uud) and a negative pion π⁻ (u`d) which being charged particles formed tracks as they travelled.
Mean lifetime of neutral kaon K⁰ (ds`) and anti neutral kaon anti-K⁰ (d`s) is $\tau \sim 120 \times 10^{-10}$ s, and $\tau \sim 120 \times 10^{-10}$ s after the interaction of **π⁻ (u`d)** and **p (uud)**, another pair of track again appeared, when K⁰ (ds`) **disintegrated** into two charged particles, e.g. into π⁺ (ud`) and π⁻ (u`d).

note:
Charged pseudoscalar mesons i.e. charged kaons, charged pions may also be produced, when high energy protons are collided with nucleons (See the topic: Production and Separation of Secondary Beams).
In a fixed-target proton accelerator, the protons are accelerated to high energies and made to collide with, say a nucleon. Due to the

collision, some of the energy of the protons is converted into secondary particles, e.g. charged kaons, charged pions, muons, neutrinos.

The secondary beams of negative kaon K^-, positive kaon K^+, negative pion π^-, positive pion π^+, muon μ^-, antimuon μ^+, muon-type neutrino ν_μ, muon-type antineutrino $\nu_\mu{}^\backprime$ may be directed to the bubble chamber filled with liquid hydrogen or with freon (CF_3Br) or with liquid-neon-hydrogen mixture and in the chamber, these particles may collide with the protons of the hydrogen atoms or with the neutrons or the protons of the nuclei of heavier atoms and interact to produce various baryons and mesons. (See the topic: Examples of Interactions).

The discovery of the charm quark resulted from the observation of massive meson states of the type $cc^\backprime$ (e.g. J/psi) in 1974
The discovery of the bottom quark followed from the detection of even heavier mesons $bb^\backprime$ (e.g. upsilon) in 1977.

note:
In the collisions of e^+e^- pairs, there ia a peak at e^+e^- cms energy of 91 GeV ($E_{cms} \sim 91$ GeV) corresponding to the Z^0 vector boson.
At e^+e^- cms energy of 91 GeV, e^+e^- pairs annihilate directly to the real Z^0 bosons which themselves are resonance states, i.e. e^+e^- pair does not annihilate to virtual photon to produce Z^0 resonance.
At e^+e^- cms energy less than say, 75 GeV (mass of Z boson ranges from 75 GeV to 100 GeV), e^+e^- pairs do not have required energy to annihilate to massive Z^0 boson, thus they annihilate to virtual photons which then decay into vector mesons which are also resonance states like Z^0 boson.

Quarkonium

Quarkonium (plural: Quarkonia) is a flavourless meson which consists of a quark and its antiquark.

Quarkonium especially means charmonium and bottomonium.

The name charmonium is used for the J/ψ meson and the other excited bound states consisting of charm quark c and anticharm quark c`.

The name bottomonium is used for the upsilon meson and the other excited bound states consisting of bottom quark b and antibottom quark b`.

J/ψ Meson

The **J/ψ (J/psi) meson (cc`)** is a charmonium/quarkonium consisting of a charm quark and an anticharm quark. **It is the only particle which has two-letter name.** Its rest mass is 3.1 GeV/c^2. It is a spin 1 particle (i.e. vector meson) with zero electric charge.

note: The J/ψ meson (cc`) having spin 1 is the first excited state of charmonium.

The **charmed eta meson η_c (cc`)** having spin 0 is the lightest charmonium state or the ground state of charmonium.

The **charmed eta meson η_c** has mass m = 2983.6 MeV and mean life time $\tau = 2.04 \times 10^{-23}$ s.

The first resonance state J/ψ or ψ (1S) has mass 3.096 GeV, whereas the second resonance state ψ (2S) has rest mass 3.686 GeV. Third, fourth, fifth resonance states ψ (3S), ψ (4S), ψ (5S) have rest mass ranging from 3.773 GeV to 4.415 GeV.

J/ψ or ψ (1S) having rest mass 3.1 GeV and ψ (2S) having rest mass 3.7 GeV were first <u>discovered</u> in 1974 in electron-positron collisions at SLAC's SPEAR (Stanford Positron Electron Accelerating Ring) which was an e^+e^- synchrotron. In this synchrotron, there was a single ring some 80 meters in diameter, in which counter-rotating beams of electrons and positrons were circulated at energy up to 3.7 GeV that is, SPEAR could accelerate e^+e^- to the cms energy of about 3.7 GeV. Later on, ψ (3S), ψ (4S), ψ (5S) (rest masses ranging from 3.773 GeV to 4.415 GeV) were also observed in e^+e^- collisions.

The tauon (1.777 GeV) had also been discovered in SLAC's SPEAR in 1975.

Today, SPEAR is used as a synchrotron radiation source.

J/ψ (3.1 GeV) was **also <u>observed</u>** in 1974 in Brookhaven alternating gradient synchrotron (AGS), where **28 GeV protons were collided on a beryllium target which resulted in the production of e^+e^- pairs.** (As e^+e^- pairs recorded in coincidence on either side of the incident proton beam axis, this implied, e^+e^- produced in pairs). It was observed that the number of e^+e^- pairs each produced with energy 3.1 GeV were **more** than the number of e^+e^- pairs each produced with energy somewhat less or more than 3.1 GeV.

This implied that the collisions of the protons into beryllium produced a resonance particle of rest mass 3.1 GeV which then decayed into e^+e^- pair.

Number of this resonance particle called J/ψ produced with energy 3.1 GeV were **more** than the number of this resonance particle produced with energy somewhat less or more than 3.1 GeV. That is, there was a resonance state of mass 3.1 GeV.

note: Resonance particle does not have definite mass, instead a distribution of mass. Thus, when J/ψ mesons (rest mass: 3.1 GeV) were produced, many of them had energy 3.1 GeV and many of them had energy somewhat less or more than 3.1 GeV. (Also see the topic: Cross-section and Resonance).

J/ψ has long life time. Its mean life time τ = 7.1 × 10⁻²¹ s which is short enough so that decay is clearly by the strong process but it is a thousand times slower than a strong decay 'ought' to be.

note: A short-lived particle has a rather indeterminate mass, spread over a wide range called the width $\Gamma = \hbar/\tau$ (measured in keV or MeV). The J/ψ meson by contrast has mass that is very precise that is, the width (Γ = 92.9 KeV) is relatively extremely small. It means, it has relatively prolonged mean lifetime τ, since mean lifetime is inversely proportional to the width.

The decay of the J/ψ (cc`) into 3 pions: **J/ψ cc`) → π⁺ (ud`) + π⁻ (u`d) + π⁰ (uu`) is OZI suppressed (but not forbidden),** just as the decay of the phi meson φ (ss`) into 3 pions: φ (ss`) → π¹ (ud`) + π⁻ (u`d) + π⁰ (uu`) is OZI suppressed. (See the topic: Examples of the Decays – Example 31).

Whereas the phi meson φ (ss`) decays into 2 Kaons: **φ (ss`) → K⁰ (ds`) + K⁰` (d`s),** J/ψ (cc`) cannot decay into 2 charged D Mesons. The mass of J/ψ (3.1 GeV) < mass of positive D mesons D⁺ (1.869 GeV) + mass of negative D meson D⁻ (1.869 GeV). Thus, the decay of the J/ψ (cc`) into 2 charged D Mesons: **J/ψ (cc`) → D⁺ (cd`) + D⁻ (c`d)** is **kinematically forbidden**.

note: Neutral kaon K^0 (ds`) and antineutral kaon $K^{0}$` (d`s) are similar to negative D meson $D^-$ (c`d) and positive D meson D^+ (cd`) but with charmed quark in D mesons replaced by strange quarks in kaons.

The branching ratio for the decay of the J/ψ meson into hadrons is 87.7 %.

The J/ψ meson can decay into hadrons through 3 gluons, through virtual photon, through '2 gluons and 1 virtual photon'.

The branching ratio for the decay of the J/ψ meson into hadrons via 3 gluons, via virtual photon, via '2 gluons and 1 virtual photon' is 64.1 %, 13.5 %, 8.8 % respectively.

The decay of the J/ψ meson into hadrons through 3 gluons: In this case, each of the charm quark and the anticharm quark emits a gluon and then the charm quark and the anticharm quark also annihilate to a gluon. These gluons decay into quark-antiquark pairs. (See the topic: Examples of the Decays – Example 31 which explains the decay of φ (ss`) through 3 gluons).

The decay of the J/ψ meson into hadrons through virtual photon: In this case, the charm and the anticharm quarks of the J/ψ meson annihilate to a virtual photon which then decays into quark-antiquark pair.

The decay of the J/ψ meson into hadrons through two gluons and one virtual photon: In this case, each of the charm quark and the anticharm quark emits a gluon and then the charm quark and the anticharm quark annihilate to a virtual photon. These gluons and the virtual photon decay into quark-antiquark pairs.

J/ψ (cc`) or ψ (1S) also decays into an electron-positron pair e^+e^- or a muon-antimuon pair $μ^+μ^-$:
J/ψ (cc`) → $e^+ + e^-$ or $μ^+ + μ^-$

The **charm quark c** and the **anticharm quark c`** of the J/ψ meson annihilate to a virtual photon which subsequently decays into an electron-positron pair **e⁺e⁻** or a muon-antimuon pair **μ⁺μ⁻**. Thus, this decay of the J/ψ meson is the electromagnetic decay.

ψ (2S) decays into ψ (1S), a positive pion π⁺ (ud`) and a negative pion π⁻ (u`d):

ψ (2S) → ψ (1S) + π⁺ (ud`) + π⁻ (u`d)

In this decay, each of the charm quark c and the anticharm quark c` of ψ (2S) emits a gluon. One gluon decays into an up quark u and an antiup quark u`. The other gluon decays into a down quark d and an antidown quark d`. These quarks and antiquarks combine to produce a positive pion π⁺ (ud`) and a negative pion π⁻ (u`d). The charm quark c and the anticharm quark c` combine to produce J/ψ (cc`) or ψ (1S). Through the emission of the two gluons, **ψ (2S) having rest mass 3.7 GeV** loses energy and transforms into **J/ψ having rest mass 3.1 GeV** which contains the same composition of the fundamental particles as **ψ (2S), i.e.** one charm quark and one anticharm quark.

To memorize D mesons, i.e. D⁺ (cd`), D⁻ (c`d), D⁰ (cu`), D⁰` (c`u), Ds⁺ (cs`), Ds⁻ (c`s), you may do this.
Write d` d u` u s` s on a paper at a gap of some cms. Then.
write c and c`alternately on the left side of each of these six. Close the brackets. Then write the symbol of corresponding D meson name on the left side of the brackets.

**Each of ψ (3S), ψ (4S), ψ (5S), whose rest masses range from 3.773 GeV to 4.415 GeV, decays into a neutral D meson D⁰ (cu`) and an antineutral D meson D⁰` (c`u),** since the rest mass of each of these higher resonance states is more than the sum (3.728 GeV) of the rest mass of the neutral D meson D$^0$ (1.864 GeV) and the rest mass of the antineutral D meson D$^{0`}$ (1.864 GeV):

ψ (3S) or ψ (4S) or ψ (5S) → D^0 (cu`) + D$^{0`}$ (c`u)

The branching ratio for this decay is 52 %.

In this decay, the charm quark of ψ (3S) **emits a gluon** and remains itself as a **charm quark c**.

The **gluon then decays into an up quark u and an antiup quark u`.**

This **antiup quark u`** combines with that **charm quark c** of ψ (3S) which had emitted the gluon and produces a **neutral D meson D$^0$ (cu`).**

The **up quark u** (produced by the decay of the gluon) combines with the **anticharm quark c`** of ψ (3S) and produces an **antineutral D meson D$^{0`}$ (c`u).**

ψ (4S), ψ (5S) may decay in the same way.

note:

Instead of the charm quark, the **anticharm quark of ψ (3S) may also emit a gluon** and would remain itself as an **anticharm quark c`.** The **gluon then decays into an up quark u and an antiup quark u`.** This **up quark u** combines with that **anticharm quark c`** of ψ (3S) which had emitted the gluon and produces an **antineutral D meson D$^{0`}$ (c`u).** The **antiup quark u`** (produced by the decay of the gluon) combines with the **charm quark c** of ψ (3S) and produces a **neutral D meson D^0 (cu`).**

Similarly, in many other strong decays, this kind of a situation may exist.

Each of ψ (3S), ψ (4S), ψ (5S) also decays into a **positive D meson D$^+$ (cd`) and a negative D meson D$^-$ (c`d),** since the rest mass of each of these higher resonance states is more than the sum (3.738 GeV) of the rest mass of the positive D meson D$^+$ (1.869 GeV) and the rest mass of the negative D meson D$^-$ (1.869 GeV):

ψ (3S) or ψ (4S) or ψ (5S) → D⁺ (cd`) + D⁻ (c`d)

The branching ratio for this decay is 41 %.

In this case, the charm quark of ψ (3S) **emits a gluon** and remains itself as a **charm quark c**.

The **gluon then decays into a down quark d and an antidown quark d`**. This **antidown quark d`** combines with that **charm quark c** of ψ (3S) which had emitted the gluon and produces a **positive D meson D⁺ (cd`).**

The **down quark d** (produced by the decay of the gluon) combines with the **anticharm quark c`** of ψ (3S) and produces a **negative D meson D⁻ (c`d).** Similarly for ψ (4S), ψ (5S).

Note:

1. neutral kaon K⁰ (ds`) and antineutral kaon K⁰` (d`s)

First symbol inside the bracket is d and d` respectively

2. neutral B meson B⁰ (db`) and an antineutral B meson B⁰` (d`b)

First symbol inside the bracket is d and d` respectively

3. neutral D meson D⁰ (cu`) and an antineutral D meson D⁰` (c`u)

First symbol inside the bracket is c and c` respectively

Upsilon (Υ) Meson

The **Upsilon (Υ) meson bb`** is a bottomonium/quarkonium consisting of a bottom quark and an antibottom quark. Its rest mass is 9.46 GeV/c². It is a spin 1 particle (i.e. vector meson) with zero electric charge.

The first resonance state Υ (1S) has mass 9.46 GeV, whereas the second, third, fourth resonance states Υ (2S), Υ (3S), Υ (4S) have rest mass 10.02 GeV, 10.36 GeV, 10.58 GeV respectively.

The upsilon meson (Υ) was first <u>discovered</u> at Fermilab near Chicago, where 400 GeV protons were collided on a beryllium or copper target which resulted in the production of μ⁺μ⁻ pairs. It was observed that the number of μ⁺μ⁻ pairs each produced with energy of about 10 GeV was **more** than the number of μ⁺μ⁻ pairs each produced with energy somewhat less or more than about 10 GeV.
This implied that the collisions of protons into beryllium or copper produced a resonance particle of rest mass about 10 GeV which then decayed into μ⁺μ⁻ pairs.

Numbers of this resonance particle (called Υ) produced with the energy 10 GeV were **more** than the number of this resonance particle produced with the energy somewhat less or more than 10 GeV. That is, there was a resonance state of mass 10 GeV.

From the graph between 'cross-section (equivalent to the number of μ⁺μ⁻ pairs produced)' versus μ⁺μ⁻ pair energy, it was observed, there was a broad peak centred around μ⁺μ⁻ pair energy of 10 GeV which implied there were two or three resonance states present with masses around 10 GeV. These states were called the upsilon.

Later on, Υ(1S), Υ (2S) were <u>observed</u> in e$^+$e$^-$ collisions at the DORIS storage ring in Hamburg and their masses were clearly determined as 9.46 GeV and 10.02 GeV. **Υ (3S) and Υ (4S) were <u>observed</u> in e$^+$e$^-$ collisions at CESR** (Cornell Electron-Positron Storage Ring), Cornell and their masses were determined as 10.36 GeV and 10.58 GeV respectively.

Υ or Υ (1S) has long lifetime. Its mean lifetime $\tau = 1.2 \times 10^{-20}$ s is greater than the mean lifetime (0.71×10^{-20} s) of J/ψ .

The branching ratio for the decay of the Υ meson into hadrons is 87.7 %.

The Υ meson can decay into hadrons through 3 gluons, through virtual photon, through '2 gluons and 1 virtual photon'.

The branching ratio for tho decay of the Υ meson into hadrons via 3 gluons, via '2 gluons and 1 virtual photon' is 81.7 %, 2.2 % respectively.

Υ (bb`) or Υ (1S) also decays into an electron-positron pair e$^+$e$^-$ or a muon-antimuon pair $\mu^+\mu^-$ or a tauon-antitauon pair $\tau^+\tau^-$:

Υ (bb`) $\rightarrow$ e$^+$ + e$^-$ or μ^+ + μ^- or τ^+ + τ^-

The **bottom quark b** and **antibotom quark b`** of the Υ (1S) meson annihilate to a virtual photon which subsequently decays into an electron-positron pair **e$^+$e$^-$** or a muon-antimuon pair **$\mu^+\mu^-$** or tauon-antitauon pair **τ⁺τ⁻**. Thus, this decay of the Υ meson is the electromagnetic decay.

Υ (2S) decays into Υ (1S), a positive pion π⁺ (ud`) and a negative pion π⁻ (u`d):

Υ (2S) → Υ (1S) + π⁺ (ud`) + π⁻ (u`d)

In this case, each of the bottom quark b and the antibottom quark b` of ψ (2S) emits a gluon. One gluon decays into an up quark u and an antiup quark u`. The other gluon decays into a down quark d and an antidown quark d`. These quarks and antiquarks combine to produce a positive pion π⁺ (ud`) and a negative pion π⁻ (u`d). The bottom quark b and the antibottom quark b` combine to produce Υ (1S).

Through the emission of the two gluons, **Υ (2S) having rest mass 10.02 GeV** loses energy and transforms into **Υ (1S) having rest mass 9.46 GeV** which contains the same composition of the fundamental particles as **Υ (2S), i.e.** one bottom quark and one antibottom quark.

To memorize B mesons, i.e. B⁺ (ub`), B⁰ (db`), Bₛ⁰ (sb`), B_c⁺ (cb`), you may do this.
Write u d s c on a paper at a gap of some cms. Then. write b`in front of each of these four. Close the brackets. Then write the symbol of corresponding B meson name on the left side of the brackets.

Υ (4S) decays into a positive B meson B⁺ (ub`) and a negative B meson B⁻ (u`b), since the rest mass (10.579 GeV) of Υ (4S) is more than the sum (10.558 GeV) of the rest mass of the positive B meson B⁺ (5.279 GeV) and the rest mass of the negative B meson B⁻ (5.279 GeV):

Υ (4S) → B⁺ (ub`) + B⁻ (u`b)

The branching ratio for this decay is 51.4 %.

In this case, the bottom quark of Υ (4S) **emits a gluon** and remains itself as a **bottom quark b**.

The **gluon then decays into an up quark u and an antiup quark u`.**

This **antiup quark u`** combines with that **bottom quark b** of $\Upsilon$ (4S) which had emitted the gluon and produces a **negative B meson B⁻ (u`b).**

The **up quark u** (produced by the decay of the gluon) combines with the **antibottom quark b`** of $\Upsilon$ (4S) and produces a **positive B meson B⁺ (ub`).**

Υ **(4S) also decays into a neutral B meson B⁰ (db`) and an antineutral B meson B⁰` (d`b), since the rest mass (10.579 GeV) of** $\Upsilon$ (4S) is more than the sum (10.558 GeV) of the rest mass of the **neutral B meson B⁰** (5.279 GeV) and the rest mass of the **antineutral B meson B⁰`** (5.279 GeV):

Υ (4S) → B⁰ (db`) + B⁰` (d`b)

The branching ratio for this decay is 48.6 %.

In this case, the bottom quark of Υ (4S) **emits a gluon** and remains itself as a **bottom quark b**.

The **gluon then decays into a down quark d and an antidown quark d`.** This **antidown quark d`** combines with that **bottom quark b** of Υ

(4S) which had emitted the gluon and produces an **antineutral B meson B$^{0^\backprime}$ (d$^\backprime$b).**

The **down quark d** (produced by the decay of the gluon) combines with the **antibottom quark b$^\backprime$ of** Υ (4S) and produces a **neutral B meson B^0 (db$^\backprime$).**

Discovery of the W and Z bosons

Mass of W boson = 80.38 GeV, Mean lifetime of W boson τ = 3.1 x 10^{-25} s, Width of W boson Γ = 2.08 GeV.

Mass of Z boson = 91.18 GeV, Mean lifetime of Z boson τ = 2.6 x 10^{-25} s, Width of Z boson Γ = 2.49 GeV.

The W and Z bosons were discovered in Super Proton Synchrotron (a proton-antiproton collider) at CERN in January 1983 and May 1983 respectively.

The W boson was discovered in Super Proton Synchrotron through the collision of proton-antiproton pairs having cms energy 540 GeV (that is, proton and antiproton each having energy 270 GeV).

The W^+ boson decayed into a **positron e^+** and a **neutrino v_e**. Very energetic positron e^+ of energy 42 GeV was observed in a surrounding electromagnetic calorimeter.

Although colliding proton-antiproton pair pp` had cms energy of 540 GeV, the entire cms energy was not converted into the resonance state as pp` pair did not annihilate to a W^+ boson. Instead, one up quark of the proton and the antidown quark of the antiproton annihilated to a W^+ boson (**u + d`** $\rightarrow$ **W^+**) which then decayed into a positron e^+ and an electron-type neutrino v_e (**W^+** $\rightarrow$ **e^+ + v_e**).

Similarly, in pp` colliders, **$W^-$** boson may be produced by the annihilation of the down quark of the proton and one antiup quark of the antiproton (**u` + d** $\rightarrow$ **W^-**) which may decay into an electron e^- and an electron-type antineutrino v_e` (**$W^-$** $\rightarrow$ **$e^-$ + $v_e$`**).

Similarly, in pp` colliders, **$Z^0$** boson may be produced by the annihilation of the up and antiup quarks or by the annihilation of the down and antidown quarks ($u + u^` → Z^0$ or $d + d^` → Z^0$). The Z^0 boson may then decay into e^+e^- pair or $\mu^+\mu^-$ pair ($Z^0 → e^+ + e^-$ or $\mu^+ + \mu^-$).

In October-December 1982, when ~ 10^9 $pp^`$ pairs were collided at the cms energy of 540 GeV (i.e. each proton-antiproton pair $pp^`$ having energy 540 GeV), 5 events of $W^+ → e^+ + v_e$ were **observed** and 52 W^+ bosons were produced, i.e. about 10% decayed into $e^+ + v_e$. Note that Br ($W^+ → e^+ + v_e$) is ~ 10.7%.

The other W^+ bosons produced by the $pp^`$ collisions, decayed into other products, e.g. $\mu^+ + v_\mu$, hadrons.

In April-May 1983, when 8×10^9 $pp^`$ were collided at the cms energy of 540 GeV, 55 events of $W^+ → e^+ + v_e$ were **observed** and ~ 550 W^+ bosons were produced.

Also, 4 events of $Z^0 → e^+ + e^-$ and 2 events of $Z^0 → \mu^+ + \mu^-$ were observed and ~ 100 Z^0 were produced.

The other Z^0 bosons produced by the $pp^`$ collisions, decayed into other products, e.g. hadrons.

The Z^0 boson resonance states were also observed in the high energy e^+e^- colliders: **the LEP collider at CERN and the SLC at Stanford**, from 1989 onwards. (See the topic: Cross-Section and Resonance).

At 540 GeV cms energy of colliding $pp^`$ pairs, the branching ratio for the process: $pp^` → W^+ → e^+ + v_e$ is ~ 5×10^{-9}, i.e. for every 10^9 $pp^`$ pairs collided, only 5 positrons e^+ and 5 neutrinos v_e were produced through the decay of the W^+ bosons.

However, the extraction of such a rare event was possible because the positron, from the decay of such a W^+ boson has very high transverse momenta (~ 40 GeV/c), and very energetic positron e^+ of energy 42

GeV from the decay of the W$^+$ boson was observed in a surrounding electromagnetic calorimeter.

note:

The different flavours $(e^-, \mu^-, \tau^-, \nu_e, \nu_\mu, \nu_\tau)$ **of leptons carry the same unit of weak charge that is, the different flavours of leptons have identical couplings to the W and Z bosons.**

For example, at LEP e^+e^- collider, the **branching ratio for the decay of the W⁺ boson into e⁺ + ν_e** is the same as the branching ratios for the decay of the W⁺ boson into $\mu^+ + \nu_\mu$ and $\tau^+ + \nu_\tau$.

Br $(W^+ \rightarrow e^+ + \nu_e)$ = 10.7%, Br $(W^+ \rightarrow \mu^+ + \nu_\mu)$ = 10.6%, Br $(W^+ \rightarrow \tau^+ + \nu_\tau)$ = 11.3%.

Similarly, the **branching ratio for the decay of the Z⁰ boson into e⁺e⁻** is the same as the branching ratios for the decay of the Z⁰ boson into $\mu^+\mu^-$ pair and $\tau^+\tau^-$ pair.

Br $(Z^0 \rightarrow e^+ + e^-)$ = Br $(Z^0 \rightarrow \mu^+ + \mu^-)$ = 3.36%, Br $(Z^0 \rightarrow \tau^+ + \tau^-)$ = 3.37%.

However, the different flavours (u, c, t, d, s, b) **of quarks don't have identical couplings to the W and Z bosons.**

That is, in contrast to the universality of the leptons couplings, the couplings of the quarks to the W and Z bosons do depend on the quark flavour.

The branching ratio for the decay of the W⁺ boson into a quark-antiquark pair is proportional to the square of the corresponding element in CKM matrix and also to the number of colour charges.

V_{ud} = 0.974	V_{us} = 0.225	V_{ub} = 0.004	u to d s b
V_{cd} = 0.225	V_{cs} = 0.986	V_{cb} = 0.041	c to d s b
V_{td} = 0.008	V_{ts} = 0.040	V_{tb} = 0.999	t to d s b

As $|V_{ud}|^2 = 0.974^2 = 0.948$ and $|V_{cs}|^2 = 0.986^2 = 0.972$ are close to unity, thus the **quark-antiquark pairs, in which the W⁺ bosons mainly decay are ud` and cs`.**

The W⁺ boson may decay into us`, ub`, cd`, cb` too but with very low probability as these decays are CKM suppressed.

Also, W⁺ boson cannot decay into td`, ts`, tb`, as the mass of a W⁺ boson is less than the mass of a top quark, thus these decays of W⁺ bosons are kinematically forbidden.

Quarks carry 3 colour charges, so **each of ud` and cs` can be of 3 types**, e.g. ud` may **have** red-antired colour or blue-antiblue colour or green-antigreen colour.

Thus, **the W⁺ bosons predominantly decay into 9 products**: e⁺ + vₑ , μ⁺ + v_μ , τ⁺ + v_τ, **ud`** with rr` or bb` or gg` colours, **cs`** with rr` or bb` or gg`colours.

Thus, Br (W⁺→ e⁺ + vₑ) = Br (W⁺ → μ⁺ + v_μ) = Br (W⁺ → τ⁺ + v_τ) is ~ 1/9 or ~11%., and Br (W⁺→ ud`) = Br (W⁺→ cs`) is ~ 3/9 or ~33%. The branching ratio for the decay into other quark-antiquark pairs: Br (W⁺→ us`), Br (W⁺→ ub`), Br (W⁺→ cd`), Br (W⁺→ cb`) is negligible. Thus, the **hadronic branching ratio of the W⁺ boson: Br (W⁺ → hadrons) ~ 0.674 or 67.4% is dominated by the CKM favoured ud` and cs` final states.**

note:

$|V_{ub}|^2 = .004^2 = 0.000016$ implies the probability with which an up quark u (charge: +2/3e) can **transform** into a bottom quark b (charge: -1/3e), by emitting a W⁺ boson or by absorbing a W⁻ boson **(u → b).** This probability is **almost zero.**

$|V_{ub}|^2 = 0.000016$ also implies the probability with which a bottom quark b (charge: -1/3e) can **transform** into an up quark u (charge: +2/3e), by emitting a W⁻ boson or by absorbing a W⁺ boson **(b → u).** This probability is **almost zero.**

$|V_{ub}|^2 = 0.000016$ also implies the probability with which an antiup quark u` (charge: -2/3e) can **transform** into an antibottom quark b` (charge:

+1/3e), by emitting a W^- boson or by absorbing a W^+ boson **(u` → b`)**. This probability is **almost zero.**

$|V_{ub}|^2 = 0.000016$ also implies the probability with which an antibottom quark b` (charge: +1/3e) can **transform** into an antiup quark u` (charge: -2/3e), by emitting a W^+ boson or by absorbing a W^- boson **(b` → u`)**. This probability is **almost zero.**

$|V_{ub}|^2 = 0.000016$ also implies the probability, with which a W^+ can **decay** into an up quark u and an antibottom quark b` **(W⁺ → ub`)**. This probability is **almost zero.**

$|V_{ub}|^2 = 0.000016$ also implies the probability with which a W^- can **decay** into an antiup quark u` and a bottom quark b **(W⁻ → u`b)**. This probability is **almost zero.**

$|V_{ub}|^2 = 0.000016$ also implies the probability, with which an up quark u and an antibottom quark b` can **annihilate** to a $W^+$ boson **(ub` → W⁺)**. This probability is **almost zero.**

$|V_{ub}|^2 = 0.000016$ also implies the probability with which an antiup quark u` and a bottom quark b can **annihilate** to a $W^-$ boson **(u`b → W⁻)**. This probability is **almost zero.**

Similarly, all the other 8 elements of CKM matrix give different probabilities for the other weak processes.

note:

The virtual W⁻ boson **may** decay into 9 products: e⁻ $\bar{\nu}_e$, μ⁻ $\bar{\nu}_\mu$, τ⁻ $\bar{\nu}_\tau$, **u`d,** u`s, u`b, c`d, c`s, c`b.

However, **during the weak decays of sigma baryon Σ, lambda baryon Λ, Xi baryon Ξ, omega baryon Ω, negative kaon K⁻, etc. (in which strangeness is increased by 1 after the decay), the virtual W⁻ boson** dominantly **decays into an antiup quark and a down quark (W⁻ → u`d) out of the 9 decay modes mentioned above.**

First, consider the decay of such a virtual W⁻ boson into **u`d,** u`s, u`b, c`d, c`s, c`b.

The decay of the virtual W⁻ boson into anti-u and d (W⁻ → u`d) is not suppressed: $|V_{ud}|^2 = 0.974^2 = 0.948$. **(note: u and d are the first, i.e. the same generation quarks).**

The decay of the virtual W⁻ boson into anti-u and s **(W⁻ → u`s) is CKM suppressed:** $|V_{us}|^2 = 0.225^2 = 0.050$,

The decay of the virtual W⁻ boson into anti-u and b **(W⁻ → u`b) is highly CKM suppressed:** $|V_{ub}|^2 = 0.004^2 = 0.000016$.

The decay of the virtual W⁻ boson into anti-c and d **(W⁻ → c`d) is CKM suppressed:** $|V_{cd}|^2 = 0.225^2 = 0.050$.

The decay of the virtual W⁻ boson into anti-c and s (W⁻ → c`s) is not suppressed: $|V_{cs}|^2 = 0.986^2 = 0.972$. **(note: c and s are the second, i.e. the same generation quarks).**

The decay of the virtual W⁻ boson into anti-c and b **(W⁻ → c`b) is highly CKM suppressed:** $|V_{cb}|^2 = 0.041^2 = 0.0016$.

Also, W⁻ boson cannot decay into anti-t and d (W⁻ → t`d), anti-t and s (W⁻ → t`s), anti-t and b (W⁻→ t`b) as the mass of a W⁻ boson is less than the mass of a top quark, thus these decays of W⁻ bosons are **kinematically forbidden.**

As the hadrons Σ, Λ, Ξ, Ω, K⁻, etc, consist of u, d and/or s quarks, so the **virtual W⁻ boson (emitted by a down or strange quark inside these hadrons)** will have energy less than the rest mass of the charm quark. Thus, the decay of **such** a virtual W⁻ boson into anti-c and s (W⁻→ c̄`s) is **kinematically forbidden,** although this decay is not CKM suppressed.

Thus, the only quark and antiquark, in which a virtual W⁻ boson may decay in each of these hadrons is anti-up and down.

To understand the decay of such a virtual W⁻ boson into e⁻ v̄ₑ`, μ⁻ v̄ᵤ`, τ⁻ v̄τ`, consider the following decays of the negative sigma baryon:

Σ⁻ (dds) → n (udd) + π⁻ (u`d) (BR: 99.848 %).

Σ⁻ (dds) → n (udd) + e⁻ + v̄ₑ` (BR: 0.1017 %).

In each of these decays, the strange quark (charge: -1/3e) emits a W⁻ boson and transforms into up quark and **Σ⁻ (dds) transforms into n (udd).**

Decay of the negative sigma baryon into the neutron, electron and electron type antineutrino is a semileptonic decay.

In this decay, **ΔS = 1** (strangeness is one less or one more after the decay).

In case of **semileptonic decay,** if ΔS = 1, the **coupling constant** α_w for the **decay of the virtual W⁻ boson** into e⁻ and v̄ₑ` (W⁻ → e⁻ v̄ₑ`) is **proportional to G sinθ₀** that is, effective coupling is proportional to G sinθ$_c$ θ$_c$ ~ 12⁰ implies G sinθ$_c$ = 0.208 G. That is, the **virtual W⁻ boson emitted by the strange quark s of Σ⁻ (dds) has much less probability to decay into e⁻ and v̄ₑ`.** As muon and tauon are much more massive than electron, the probability of such a W⁻ boson to decay into μ⁻ and v̄ᵤ` or into τ⁻ and v̄τ` is almost zero.

Thus, in the decay of the <u>negative sigma baryon</u> (in which strangeness is increased by 1 after the decay), out of nine possible decay modes for the decay of W⁻ boson, the only dominant decay mode is antiup and down quark (W⁻ → u`d). These antiup and down quarks produced by the decay of the virtual W⁻ boson combine to produce a negative pion: Σ⁻ (dds) → n (udd) + π⁻ (u`d) (BR: 99.848 %).

Similarly, in the decays of <u>Λ, Ξ, Ω, K⁻, etc.,</u> the only dominant decay mode for the **W⁻** is antiup and down quark **(W⁻ → u`d).**

The decay of the virtual W⁻ boson into an electron e⁻ and an electron-type antineutrino v_e` (W⁻ → e⁻ v_e`) depends upon the type of the decay process as explained below:

The **weak charge g_w** at low energy is calculated from the equation: $g_w{}^2/8M_W = G/2^{1/2}$, where G = 1.166 × 10⁻⁵ GeV⁻² is Fermi constant and M_W = 80.38 MeV is the mass of the W boson. These values imply **g_w =** **0.65.**

The **weak coupling constant** at low energy α_w = g_w²/4π = 0.65²/4π = 1/29.5

The **coupling constant** α_w for the weak interaction specifies the strength with which a particle having weak charge **emits** or **absorbs** a W boson. This also implies the strength with which a W boson **decays** into particles having weak charges.

$g_w{}^2/8M_W = G/2^{1/2}$ implies $g_w{}^2$ is proportional to Fermi constant G and as α_w = g_w²/4π, **thus α_w is also proportional to Fermi constant G.**

Now consider the following examples:

1. μ⁻ → e⁻ + v_e` + v_μ (BR: 100 %); leptonic decay; μ⁻ never decays into π⁻ (u`d) + v_μ

In this decay, the muon (charge: -1e) **emits a W⁻ boson** and transforms into a muon-type neutrino.

2. n (udd) → p (uud) + e⁻ + v$_e$`(BR: ~ 100 %); semileptonic decay, ΔS = 0

In this decay, the down quark (charge: -1/3e) **emits a W⁻ boson** and transforms into an up quark and **n (udd) transforms into p (uud).**

3. Σ⁻ (dds) → n (udd) + e⁻ + v$_e$` (BR: 0.1017 %); semileptonic decay, ΔS = 1

In this decay, the strange quark (charge: -1/3e) **emits a W⁻ boson** and transforms into an up quark and **Σ⁻ (dds) transforms into n (udd).**

Decay of the muon into the electron, electron type antineutrino and muon type neutrino is a leptonic decay: μ⁻ → e⁻ + v$_e$` + v$_\mu$.

In case of **leptonic decay,** the **coupling constant α$_w$** for the decay of the virtual W⁻ boson into an electron e⁻ and an electron-type antineutrino v$_e$` (W⁻ → e⁻ v$_e$`) is indeed **proportional to Fermi constant G.** Thus, the virtual W⁻ boson **emitted** by a muon μ⁻ always decays into e⁻ and v$_e$`.

Decay of the neutron into the proton, electron and electron type antineutrino is a semileptonic decay: n (udd) → p (uud) + e⁻ + v$_e$`.

In this decay, **ΔS = 0** (strangeness is the same before as well as after the decay),

In case of **semileptonic decay,** involving u and d quarks, **if ΔS = 0,** the **coupling constant α$_w$** for the decay of the virtual W⁻ boson into e⁻ and v$_e$` (W⁻ → e⁻ v$_e$`) is **proportional to G cosθ$_c$** (θ$_c$ is called Cabibbo angle) that is, effective coupling is proportional to G cosθ$_c$. Experiments give θ$_c$ ~ 12⁰ so that G cosθ$_c$ = G cos12⁰ = 0.978 G which is almost equal to G. Thus, the virtual W⁻ boson **emitted** by the down quark d of n (udd) almost always decays into e⁻ and v$_e$`.

Decay of the negative sigma baryon into the neutron, electron and electron type antineutrino is a semileptonic decay: Σ^- (dds) $\rightarrow$ n (udd) + e$^-$ + v_e`. In this decay, ΔS = 1 (strangeness is one less or one more after the decay).

In case of **semileptonic decay,** if ΔS = 1, the **coupling constant** α_w for the decay of the virtual W$^-$ boson into e$^-$ and v_e` (W$^-$ $\rightarrow$ e$^-$ $v_e$`) is **proportional to G sinθ_c**.

Accelerators and Colliders

The 3.2km long linac (electron linear accelerator) at the SLAC (Stanford Linear Accelerator Center) is the largest electron linear accelerator and can accelerate <u>each electron up to 50 GeV.</u>

All proton accelerators and many electron accelerators are circular which are called cyclic accelerators or synchrotrons. In synchrotrons, the particles are accelerated once or more per revolution by radio frequency cavities.

Tevatron, the Fermilab <u>proton-antiproton synchrotron</u> was 1 km in radius and could accelerate <u>each proton and each antiproton up to 1000 GeV.</u> The top quark was discovered in Tevatron in 1995.

SPS (Super Proton Synchrotron) was also a proton-antiproton synchrotron in which W, Z bosons had been discovered in 1983.

Under the circular acceleration, an electron emits synchrotron radiation. The energy radiating (due to the synchrotron radiation) per particle per turn is 10^{13} times more for electron than proton. For an electron of energy 20 GeV circulating in a ring of radius 1 Km, this energy loss is 16 MeV per turn. Thus, circular e^+e^- colliders cannot be developed for the higher energies. Instead, linear e^+e^- colliders are used to obtain higher energy.

SLC (SLAC Linear Collider) at the Stanford is the first <u>linear e^+e^- collider</u> and can accelerate <u>each electron and each positron up to 50 GeV</u> and then the two beams at the end of the accelerator are separated by magnetic elements. Then the **counter-rotating beams of e^+e^- are made to collide head-on.** That is, the beams of electrons and positrons move in circular arcs in the opposite directions and then are made to collide head-on.

LEP collider (Large Electron Positron Collider) at CERN was a circular e⁺e⁻ collider with circumference of 27 kms built in a tunnel 100 meter underground. It operated during 1989 to 2000. Later on, LHC used the LEP tunnel. LEP produced Z^0 boson in 1989. **LEP collider energy eventually reached 209 GeV** at the end in 2000.
The older circular collider SPS accelerated electrons and positrons to the velocity nearly equal to that of light. These were then injected into LEP, where they were further accelerated to higher energies and further focused to bunches.

ϒ (3S) and ϒ (4S) had been observed at CESR (Cornell Electron-Positron Storage Ring), Cornell and their masses were determined as 10.36 GeV and 10.58 GeV respectively.

CESR (Cornell Electron-Positron Storage Ring): It consists of a linac, synchrotron and storage ring. **The electrons and the positrons are first accelerated in a linac** to the cms energy of about 150 MeV. Then these e⁺e⁻ are **injected** into the synchrotron and are accelerated to the cms energy of 5 GeV. Then these e⁺e⁻ are **transferred** to the storage ring and are accelerated to the cms energy of 9 GeV to 12 GeV and then the **counter-rotating beams of stored e⁺e⁻ are made to collide at discrete locations.** The resulting interactions are then analysed in surrounding particle detectors. There are various types of particle detectors, e.g. bubble chambers, scintillating counters, Cerenkov counters, solid state counters, shower detectors and calorimeters, electromagnetic shower detectors, hadron–shower calorimeters.

Like synchrotron, in the storage ring, particles travel in a circular orbit in vacuum under the influence of a magnetic field. In synchrotron, e^+e^- spends less than one hundredth of a second (during which, they can be accelerated by many GeV), but once e^+e^- are transferred to the storage ring, they are kept circulating there for many hours. Storage rings most commonly store electrons, positrons, or protons.

LHC (Large Hadron Collider) is a proton proton collider which can accelerate protons to the cms energy of 13000 GeV, i.e. each proton can be accelerated to 6500 GeV.

note: Synchrotron radiation is the electromagnetic radiation, emitted by electron, positron, etc. which are being accelerated radially.

Bremsstrahlung is a phenomena, in which an electron undergoing deceleration in dielectric medium (e.g, in the bubble chamber) emits photons or electromagnetic radiation.

Production and Separaton of Secondary Beams

In colliding-beam machines, the two beams of particles rotating in opposite directions are collided, so that virtually all the energy is available for a new-particle production, whereas **in a fixed-target proton accelerator, the protons are accelerated to high energies and made to collide with, say a nucleon. Due to the collision, some of the energy of the protons is converted into secondary particles, e.g. charged kaons, charged pions, muons, neutrinos.** Protons release energy in the form of gluons which then decay into quark and antiquark pairs.

{p (uud) → π^+ (ud`) + n (udd),  p (uud) → $\pi^-$ (u`d) + Δ^{++} (uuu), π^+ (ud`)
→ μ^+ + v_μ, π^- (u`d) → $\mu^-$ + $v_\mu$`
p (uud) → K^+ (us`) + $\Sigma^0$ (uds), p (uud) → $K^+$ (us`) + K^- (u`s) + p (uud)}

As a secondary beam contains several types of particles, separators are used to select the type of particle required. At energy below a few GeV, the separator **consists of** two parallel plates with a high potential between them. At energies above a few GeV, radio frequency separators are used.

The difference in angular deflection of different high energy particles passing through these separators separates them from one another. In this way, beams of, e.g. **negative kaon K^-, positive kaon K^+, negative pion π^-, positive pion π^+** can be obtained separately.

To obtain the beams of muons and neutrinos, the beams of secondary kaons and pions are made to pass through a decay tunnel which is hundreds of meters long. In the decay tunnel, a **fraction of kaons and pions decays into muons and neutrinos** (π^+, K^+ → μ^+ + v_μ and π^-, K^- → μ^- + v`$_\mu$).

To obtain a neutrino beam, muons are separated by a thick iron and earth shield. Muons cannot penetrate this thick iron and earth shield but neutrinos do. The decay of kaons produce high energy neutrinos and decay of pions produce low energy neutrinos.

The secondary beams of negative kaon K^-, positive kaon K^+, negative pion π^-, positive pion π^+, muon μ^-, antimuon μ^+, muon-type neutrino v_μ, muon-type antineutrino v_μ` may be **directed** to the bubble chamber filled with **liquid hydrogen** or with **freon** (CF_3Br) or with **liquid-neon-hydrogen mixture** and in the chamber, these particles may collide with the protons of the hydrogen atoms or with the neutrons or the protons of the nuclei of heavier atoms and interact to produce various baryons and mesons.

Examples of the Interactions

1. K^- (u`s) + p (uud) → $\Lambda^0$ (uds) + $\pi^0$ (uu`)
2. K^- (u`s) + p (uud) → $\Sigma^+$ (uus) + $\pi^-$ (u`d)
3. K^- (u`s) + p (uud) → $\Sigma^-$ (dds) + $\pi^+$ (ud`)
4. K^- (u`s) + p (uud) → n (udd) + $K^{0`}$ (d`s)
5. K^- (u`s) + p (uud) → $\Omega^-$ (sss) + $K^+$ (us`) + K^0 (ds`)
6. π^- (u`d) + p (uud) → $\Lambda^0$ (uds) + $K^0$ (ds`)
7. π^- (u`d) + p (uud) → $\Sigma^0$ (uds) + $K^0$ (ds`)
8. π^- (u`d) + p (uud) → n (udd) + $\pi^0$ (uu`)
9. π^+ (ud`) + p (uud) → Δ^{++} (uuu)
10. π^+ (ud`) + p (uud) → $\Sigma^+$ (uus) + $K^+$ (us`)
11. π^+ (ud`) + p (uud) → $K^+$ (us`) + $K^{0`}$ (d`s) + p (uud)
12. K^+ (us`) + p (uud) → $\Delta^{++}$ (uuu) + $K^0$ (ds`)
13. $K^{0`}$ (d`s) + p (uud) → Λ^0 (uds) + π^+ (ud`)
14. v_μ + p → μ^- + D^+ (cd`) + p
15. Charged current weak interaction: v_μ + n (udd) → μ^- + p (uud)
16. Neutral current weak interaction: v_μ` + $e^-$ → $v_\mu$` + e^-

1. An incoming negative kaon K^- (u`s) <u>collides</u> with a proton p (uud) in the bubble chamber and <u>produces</u> a lambda baryon $\Lambda^0$ (uds) and a neutral pion $\pi^0$ (uu`):

K^- (u`s) + p (uud) → $\Lambda^0$ (uds) + $\pi^0$ (uu`)

This is a strong interaction as strangeness is conserved (strangeness is S = -1 before as well as after the interaction). Here, **one up quark of the proton and the antiup quark of the negative kaon annihilate to a gluon which** then decay into an **up quark u** and an **antiup quark u`**. This **antiup quark u`** combines with the surviving up quark of the proton to produce a **neutral pion π^0 (uu`)**.

The **up quark u** (produced by the decay of the gluon) combines with the down quark of the proton and the strange quark of the negative kaon to produce a **lambda baryon Λ^0 (uds).**

The neutral pion π^0 ($\tau = 8.52 \times 10^{-17}$ s) decays into 2 photons (2γ), then **γ-rays convert into e⁺e⁻ pairs in traversing the liquid hydrogen creating the electromagnetic shower.** The pair appears straight at the interaction point because neutral pion will travel almost zero distance in a lifetime of 8.5×10^{-17} s, even if its velocity is nearly equal to that of light.

The lambda baryon Λ^0 decays into a proton p (uud) and a negative pion π^- (u`d) after travelling some distance a few centimeters in its lifetime τ = $2.6 \times 10^{-10}$ s: **$\Lambda^0$ (uds) → p (uud) + $\pi^-$ (u`d).**

(Already explained in the topic: Examples of The Decays - example 4)

2. An incoming negative kaon K⁻ (u`s) <u>collides</u> with a proton p (uud) in the bubble chamber and <u>produces</u> a positive sigma baryon Σ⁺ (uus) and a negative pion π⁻ (u`d):

K⁻ (u`s) + p (uud) → Σ⁺ (uus) + π⁻ (u`d)

This is a strong interaction as strangeness is conserved.

Here, **one up quark of the proton and the antiup quark of the negative kaon annihilate to a gluon which** then decays into an **up quark u** and an **antiup quark u`**. This **antiup quark u`** combines with the down quark of the proton to produce a **negative pion π⁻ (u`d).**

The **up quark u** (produced by the decay of the gluon) combines with the surviving up quark of the proton and the strange quark of the negative kaon to produce a **positive sigma baryon Σ⁺ (uus).**

3. An incoming negative kaon K⁻ (u`s) collides with a proton p (uud) in the bubble chamber and produces a negative sigma baryon Σ⁻ (dds) and a positive pion π⁺ (ud`):

K⁻ (u`s) + p (uud) → Σ⁻ (dds) + π⁺ (ud`)

This is a strong interaction as strangeness is conserved.

Here, **one up quark of the proton and the antiup quark of the negative kaon annihilate to a gluon which** then decays into a **down quark d** and an **antidown quark d`**. This **antidown quark d`** combines with the surviving up quark of the proton to produce a **positive pion π⁺ (ud`).**

The **down quark d** (produced by the decay of the gluon) combines with the down quark of the proton and the strange quark of the negative kaon to produce a **negative sigma baryon Σ⁻ (dds).**

The negative sigma baryon Σ⁻ decays into a neutron n (udd) and a negative pion π⁻ (u`d) after travelling some distance a few centimeters in its lifetime τ = 1.479 × 10⁻¹⁰ s: **Σ⁻ (dds) → n (udd) + π⁻ (u`d).**

Already explained in the topic: Examples of The Decays - example 17)

4. An incoming negative kaon K⁻ (u`s) collides with a proton p (uud) in the bubble chamber and produces a neutron n (udd) and an antineutral kaon K⁰` (d`s):

K⁻ (u`s) + p (uud) → n (udd) + K⁰` (d`s)

This is a strong interaction as strangeness is conserved.

Here, **one up quark of the proton and the antiup quark of the negative kaon annihilate to a gluon which** then decays into a **down quark d** and an **antidown quark d`**. This **antidown quark d`** combines with the strange quark of the negative kaon to produce an **antineutral kaon K⁰` (d`s).**

The **down quark d** (produced by the decay of the gluon) combines with the down quark and the surviving up quark of the proton to produce a **neutron n (udd).**

5. An incoming negative kaon K⁻ (u`s) collides with a proton p (uud) in the bubble chamber and produces a negative omega baryon Ω⁻ (sss), a positive kaon K⁺ (us`) and a neutral kaon K⁰ (ds`)

K⁻ (u`s) + p (uud) → Ω⁻ (sss) + K⁺ (us`) + K⁰ (ds`)

(Already explained in the topic: Omega Baryons)

6. An incoming negative pion π⁻ (u`d) collides with a proton p (uud) in the bubble chamber and produces a lambda baryon Λ⁰ (uds) and a neutral kaon K⁰ (ds`):

π⁻ (u`d) + p (uud) → Λ⁰ (uds) + K⁰ (ds`)

This is a strong interaction as strangeness is conserved.

Here, **one up quark of the proton and the antiup quark of the negative pion annihilate to a gluon which** then materializes into a **strange quark s** and an **antistrange quark s`**. This **antistrange quark s`** combines with the down quark of the negative pion to produce a **neutral kaon K⁰ (ds`).**

The **strange quark s** (produced by the decay of the gluon) combines with the down quark and the surviving up quark of the proton to produce a **lambda baryon Λ⁰ (uds).**

7. An incoming negative pion π⁻ (u`d) collides with a proton p (uud) in the bubble chamber and produces a neutral sigma baryon Σ⁰ (uds) and a neutral kaon K⁰ (ds`):

π⁻ (u`d) + p (uud) → Σ⁰ (uds) + K⁰ (ds`)

(This interaction may occur in the same way as π⁻ (u`d) + p (uud) → Λ⁰ (uds) + K⁰ (ds`. The difference is in the mass: the lambda baryon Λ⁰ has rest mass 1116 MeV and the neutral sigma baryon has rest mass 1193 MeV).

Σ⁰ (uds) decays into Λ⁰ (uds) + γ.

8. An incoming negative pion π⁻ (u`d) collides with a proton p (uud) in the bubble chamber and produces a neutron n (udd) and a neutral pion π⁰ (uu`):

π⁻ (u`d) + p (uud) → n (udd) + π⁰ (uu`)

This is a strong interaction as strangeness is conserved (strangeness is S = 0 before as well as after the interaction).

Here, **one up quark of the proton and the antiup quark of the negative pion annihilate to a gluon which** then materializes into an **up quark u** and an **antiup quark u`**. This **antiup quark u`** combines with the surving up quark u of the proton to produce a **neutral pion π⁰ (uu`).**

The **up quark u** (produced by the decay of the gluon) combines with the down quark of the proton and the down quark of the negative pion to produce a **neutron n (udd).**

The neutral pion π⁰ (τ = 8.52 × 10⁻¹⁷ s) decays into 2 photons (2γ), then **γ-rays convert into e⁺e⁻ pairs in traversing the liquid hydrogen creating the electromagnetic shower**

9. An incoming positive pion π⁺ (ud`) collides with a proton p (uud) in the bubble chamber and produces a double positive delta baryon Δ⁺⁺ (uuu):

π⁺ (ud`) + p (uud) → Δ⁺⁺ (uuu)

This is a strong interaction as strangeness is conserved which is zero before as well as after the interaction.

Here, **the down quark of the proton and the antidown quark of the positive pion annihilate to a gluon which** is absorbed by an up quark. The up quark of the positive pion and the two up quarks of the proton combine to produce a **double positive delta baryon Δ⁺⁺ (uuu).**

note: The **double positive delta baryon Δ⁺⁺ (uuu)** was discovered in cloud chamber in 1952, when positive pions were incident on proton target. **π⁺ (ud`) + p (uud) → Δ⁺⁺ (uuu)**.

Double positive delta baryon Δ⁺⁺ (uuu) was the first of nearly one hundred resonance states of baryons and mesons discovered.

The double-positive delta baryon Δ⁺⁺ (uuu) rapidly decays into a proton p (uud) and a positive pion π⁺ (ud`):

Δ⁺⁺ (uuu) → p (uud) + π⁺ (ud`)

(Already explained in the topic: Examples of The Decays - example 8)

10. An incoming positive pion π⁺ (ud`) collides with a proton p (uud) in the bubble chamber and produces a positive sigma baryon Σ⁺ (uus) and a positive kaon K⁺ (us`):

π⁺ (ud`) + p (uud) → Σ⁺ (uus) + K⁺ (us`)

This is a strong interaction as strangeness is conserved.

Here, **the down quark of the proton and the antidown quark of the positive pion annihilate to a gluon which** then decays into a **strange quark s** and an **antistrange quark s`**. This **antistrange quark s`** combines with the up quark of the positive pion to produce a **positive kaon K⁺ (us`).**

The **strange quark s** (produced by the decay of the gluon) combines with the two up quarks of the proton to produce a **positive sigma baryon Σ⁺ (uus).**

11. An incoming positive pion π⁺ (ud`) collides with a proton p (udd) in the bubble chamber and produces a positive kaon K⁺ (us`) and an antineutral kaon K⁰` (d`s):

π⁺ (ud`) + p (uud) → K⁺ (us`) + K⁰` (d`s) + p (uud)

This is a strong interaction as strangeness is conserved.

Here, the up quark of the positive pion **emits a gluon** and remains itself as an **up quark u.** The gluon then materializes into a **strange quark s** and an **antistrange quark s`.** This **antistrange quark s`** combines with that **up quark u** of the positive pion which had emitted gluon and produces a **positive kaon K⁺ (us`).**

The **strange quark s** (produced by the decay of the gluon) combines with the **antidown quark d`** of the positive pion and produce**s an antineutral kaon K⁰` (d`s).**

12. An incoming positive kaon K⁺ (us`) collides with a proton p (uud) and produces a double positive delta baryon Δ⁺⁺ (uuu) and a neutral kaon K⁰ (ds`):

K⁺ (us`) + p (uud) → Δ⁺⁺ (uuu) + K⁰ (ds`)

This is a strong interaction as strangeness is conserved.

Here, the up quark of the positive kaon or one of the two up quarks of the proton **emits a gluon** and the up quark remains itself as an up quark. The gluon then materializes into a **down quark d** and an **antidown quark d`.** This **down quark d** combines with the antistrange quark of the positive kaon to produce a **neutral kaon K⁰ (ds`).**

The **antidown quark d`** (produced by the decay of the gluon) **and the** down quark of the proton annihilate to a gluon.

This gluon is then absorbed by the up quark of the positive kaon or by one of the two up quarks of the proton. The up quark of the positive kaon and the two up quarks of the proton recombinc to produce a **double positive delta baryon Δ⁺⁺ (uuu).**

13. An incoming antineutral kaon $K^{0`}$ (d`s) collides with a proton p (uud) in the bubble chamber and produces a lambda baryon Λ^0 (uds) and a positive pion π^+ (ud`):

$K^{0`}$ (d`s) + p (uud) $\rightarrow$ Λ^0 (uds) + π^+ (ud`)

This is a strong interaction as strangeness is conserved.

Here, the **down quark of the proton and the antidown quark of the antineutral kaon annihilate to a gluon which** then decays into a **down quark d** and an **antidown quark d`**. This **antidown quark d`** combines with one of the two up quarks of the proton to produce a **positive pion π^+ (ud`).**

The **down quark d** (produced by the decay of the gluon) combines with another up quark of the proton and the strange quark of the antineutral kaon to produce a **lambda baryon Λ^0 (uds).**

14. When hydrogen bubble chamber BEBC (Big European Bubble Chamber) filled with superheated liquid hydrogen was exposed to a muon-type neutrino beam v_μ at the CERN SPS, the collision between the muon-type neutrino v_μ and proton (of hydrogen atom) produced positive D meson:

$v_\mu + p \rightarrow \mu^- + D^+$ (cd`) + p

This is a weak interaction as charmness is not conserved. Charmness is C = 0 and +1 before and after the interaction respectively.

The muon-type neutrino v_μ emitted a W^+ boson and transformed into a **muon μ^-.** The W^+ boson then decayed into a **charm quark c** (charge: +2/3e) and an **antidown quark d`** (charge: +1/3e) which combined to produce a **positive D meson $D^+$ (cd`).**

15. Charged current weak interaction: v_μ + n (udd) $\rightarrow$ μ^- + p (uud)

An incoming muon-type neutrino v_μ enters Gargamelle bubble chamber (filled with Freon: CF_3Br) from the left and <u>collides</u> with a neutron n (udd) in the bubble chamber and <u>produces</u> a muon μ^- and proton p (uud).

The muon leaves the chamber at the **right**. During this collision, the **remaining energy of the neutrino produces pions and other hadrons.** The neutral pions decay to y-rays which in turn convert into e^+e^- pairs in the heavy liquid, creating the electromagnetic shower.

note: Although, in this interaction strangeness is conserved (which is zero before as well as after the interaction), this cannot be a strong interaction, since in this interaction neutrino is involved which does not carry strong charge and therefore cannot emit a gluon. Instead muon-type neutrino v_μ with zero electric charge emits one unit of positive electric charge by emitting a W^+ boson and transforms into the corresponding charged lepton, i.e. **muon μ^-** with electric charge: -1e. Then, one of the two down quarks d (charge: -1/3e) of the neutron (udd) absorbs that W^+ boson and transforms into an up quark u (charge: +2/3e) and the neutron transforms into a proton **p (uud).**

Thus, this is a charged current weak interaction as it is mediated by a charged boson W^+.

note:

Gargamelle was a heavy liquid bubble chamber which operated in CERN between 1970 and 1979

BEBC had been in use during 1973 to 1984. Charged particles passing through it would leave trails of bubbles.

16. Neutral current weak interaction: $v_\mu` + e^- \rightarrow v_\mu` + e^-$

First neutral current weak process ($v_\mu` + e^- \rightarrow v_\mu` + e^-$) was observed in 1973 in Gargamelle bubble chamber at CERN, filled with Freon.

The muon-type antineutrino $v`_\mu$ **beam** consisting of one billion muon-type antineutrinos $v_\mu`$ per bunch obtained from the decay of the negative pions in flight ($\pi^- \rightarrow \mu^- + v_\mu`$) was **directed** to Gargamelle bubble chamber. A single electron was also **directed** to Gargamelle bubble chamber. The electron underwent bremsstrahlung and pair production ($v_\mu` + e^-$) was also observed.

In this interaction, the muon-type antineutrino $v`_\mu$ emits a $Z^0$ boson and remains itself as a muon-type antineutrino $v`_\mu$. The electron e^- then absorbs this Z^0 boson and remains itself as an electron e^-. Thus, this is a neutral current weak interaction as it is mediated by a neutral boson Z^0. In this interaction, there is an exchange of Z^0 boson between a first generation lepton (e^-) and a second generation antilepton ($v`_\mu$), thus this interaction occurs rarely.

note: Most neutral processes are obscured by electromagnetic ones. For example, the interaction: $e^- + e^+ \rightarrow \mu^- + \mu^+$ can occur either by a virtual photon γ or by a virtual Z^0 boson that is, e^+e^- pair annihilates to a virtual photon or a Z^0 boson which subsequently decays into $\mu^+\mu^-$ pair. At low energies, the electromagnetic decay dominates. Thus, **neutrino scaterring was originally used to confirm the existence of the neutral current weak interaction.** Neutrinos have no electromagnetic coupling as they do not have electric charges, so the weak effects are not obscured.

17. Production of muon-antimuon pair $\mu^+\mu^-$ during the collision of a pion (ud` or u`d) and a nucleon (uud or udd)

This is an electromagnetic interaction. In this case, the antiquark of the pion (ud` or u`d) and a quark of the nucleon (neutron or proton) annihilate to a virtual photon that decays into muon-antimuon pair $\mu^+\mu^-$.

Neutral K Meson

Production of the Neutral K Meson

An incoming negative pion π⁻ (u`d) may collide with a proton p (uud) in the bubble chamber and produce a lambda baryon Λ⁰ (uds) and a neutral kaon K⁰ (ds`):

π⁻ (u`d) + p (uud) → Λ⁰ (uds) + **K⁰ (ds`)**   (strong interaction)

(Already explained in the topic: Examples of Interactions – example 6)

An incoming positive kaon K⁺ (us`) may collide with a proton p (uud) in the bubble chamber and produce a double positive delta baryon Δ⁺⁺ (uuu) and a neutral kaon K⁰ (ds`):.

K⁺ (us`) + p (uud) → Δ⁺⁺ (uuu) + **K⁰ (ds`)**   (strong interaction)

(Already explained in the topic: Examples of Interactions – example 12)

An incoming positive pion π⁺ (ud`) may collide with a proton p (udd) in the bubble chamber and produce a positive kaon K⁺ (us`) and an antineutral kaon K⁰` (d`s):

π⁺ (ud`) + p (uud) → K⁺ (us`) + **K⁰` (d`s)** + p (uud)   (strong interaction)

(Already explained in the topic: Examples of Interactions – example 11)

An incoming negative kaon K⁻ (u`s) may collide with a proton p (uud) in the bubble chamber and produce a neutron n (udd) and an antineutral kaon K⁰` (d`s):

K⁻ (u`s) + p (uud) → n (udd) + **K⁰` (d`s)**   (strong interaction)

(Already explained in the topic: Examples of Interactions – example 4)

The neutral kaon and the antineutral kaon both have zero electric charge and the same mass. However, **they are distinguished through their strong interactions** that is,

An incoming antineutral kaon $K^{0`}$ (d`s) may collide with a proton p (uud) in the bubble chamber and produce a lambda baryon Λ^0 (uds) and a positive pion π^+ (ud`):

$K^{0`}$ (d`s) + p (uud) $\rightarrow \Lambda^0$ (uds) + π^+ (ud`)

(Already explained in the topic: Examples of Interactions – example 13)

In this interaction, strangeness is conserved (S = - 1 before as well as after the interaction).

However, the strong interaction: K^0 (ds`) + p (uud) $\rightarrow \Lambda^0$ (uds) + $\pi^+$ (ud`) cannot occur as strangeness will not be conserved (S = +1 and -1 before and after the interaction).

K^0 - $K^{0`}$ (neutral kaon – antineutral kaon) **Mixing**

An initially pure beam of neutral kaon K^0 (ds`) transforms into its antiparticle, i.e. antineutral kaon $K^{0`}$ (d`s) while propagating in space. Then antineutral kaon $K^{0`}$ (d`s) transforms back into neutral kaon $K^0$ (ds`) and so on. This is called Neutral kaon Oscillation.

The neutral kaon K^0 (ds`) can transform into the antineutral kaon $K^{0`}$ (d`s) while propagating in space as follows:

In this transformation, strangeness is not conserved as strangeness of the neutral kaon K^0 (ds`) is S = + 1 and strangeness of the antineutral kaon $K^{0`}$ (d`s) is S = - 1, thus this transformation is possible through the weak process.

The **down quark d (charge: -1/3e)** of the neutral kaon K^0 (ds`) **emits a $W^-$ boson** and transforms into an up quark u (charge: +2/3e). This up quark u (charge: +2/3e) and the antistrange quark s` (charge: +1/3e) of the neutral kaon K^0 (ds`) **annihilate to a W^+ boson**.

Thus, in this way, the down quark d and the antistrange quark s`of the neutral kaon $K^0$ (ds`) **radiate two W bosons** and **exchange an up quark** between them.

The **W^+ boson decays** into an **antidown quark d`** (charge: +1/3e) and an up quark u (charge: +2/3e). This up quark u (charge:+2/3e) **absorbs W^- boson** and transforms into a **strange quark s** (charge: -1/3e).

Thus, the two W bosons **exchange an up quark** between them.

The **antidown quark d`** and the **strange quark s** combine to produce an **antineutral kaon $K^{0`}$ (d`s)**.

Similarly, the antineutral kaon $K^{0`}$ (d`s) can transform into the neutral kaon K^0 (ds`) as follows:

The **strange quark s (charge: -1/3e)** of the antineutral kaon $K^{0`}$ (d`s) **emits a W^- boson** and transforms into an up quark u (charge: +2/3e). This up quark u (charge: +2/3e) and the antidown quark d` (charge: +1/3e) of the antineutral kaon $K^{0`}$ (d`s) **annihilate to a W^+ boson**.

Thus, in this way, the strange quark s and the antidown quark d`of the antineutral kaon $K^{0`}$ (d`s) **radiate two W bosons** and **exchange an up quark** between them.

The **W^+ boson decays** into an **antistrange quark s`** (charge: +1/3e) and an up quark u (charge: +2/3e). This up quark u (charge: +2/3e) **absorbs W^- boson** and transforms into a **down quark d** (charge: -1/3e). Thus, the two W bosons **exchange an up quark** between them.

The **down quark d** and the **antistrange quark s`** combine to produce a **neutral kaon $K^0$ (ds`)**.

In this example, K^0 - $K^{0`}$ Mixing is due to the exchange of up (charge: +2/3e) quarks. $K^0$ - $K^{0`}$ Mixing may be due to the exchange of charm (charge: +2/3e) or top (charge: +2/3e) quarks too.

Short-lived Kaon K_S and long-lived Kaon K_L

Due to the neutral kaon oscillation i.e. due to the mixing of K^0 $K^{0`}$ states, two other particles are produced. One is the sum of these two states and is called short-lived Kaon K_S and the other is the difference of these two states and is called long-lived Kaon K_L.

K_S = $(K^0 + K^{`0})/2^{1/2} \rightarrow 2\,\pi$  and  $K_L$ = $(K^0 - K^{0`})/2^{1/2} \rightarrow 3\pi$

K_S decays into 2 pions ($\pi^+ \pi^-$ or $\pi^0 \pi^0$) in 0.89×10^{-10} s.

K_L decays into 3 pions ($\pi^+ \pi^- \pi^0$) in 0.52×10^{-7} s, i.e. **585** times slower than short-lived Kaon K_S.

Thus, K_S are K_L are distinguished by their decay modes.

CP violation in the Kaon system

Whereas 32% of all long-lived kaons K_L decay by the 3 π mode, 40.55% of all long-lived kaons K_L decay by leptonic modes.

The leptonic decay modes of K_L ($K_L \rightarrow \pi^- + e^+ + v_e$ and $K_L \rightarrow \pi^+ + e^- + v_e^{`}$) violate CP invariance, i.e. there is small charge asymmetry.

note:

The electron-type neutrino v_e being a neutrino has LH helicity and the electron-type antineutrino $v_e^{`}$ being an antineutrino has RH helicity.

These decays are weak decays, i.e. through the W bosons as neutrino and antineutrino are involved.

If we apply **parity operation** to the decay: $K_L \rightarrow \pi^- + e^+ + v_e$ (where v_e is **LH**), the decay process will become $K_L \rightarrow \pi^- + e^+ + v_e$ (where v_e is **RH**)

This can be written as: $P \mid K_L \rightarrow \pi^- + e^+ + v_{e\,LH} > = K_L \rightarrow \pi^- + e^+ + v_{e\,RH}$

Now, if we apply **charge conjugation operation** to the decay: $K_L \rightarrow \pi^- + e^+ + v_{e\,RH}$, then the decay process will become $K_L \rightarrow \pi^+ + e^- + v_e\grave{}_{RH}$

This can be written as: $C \mid K_L \rightarrow \pi^- + e^+ + v_{e\,RH} > = \mid K_L \rightarrow \pi^+ + e^- + v_e\grave{}_{RH} >$

That is, **if we apply P operation and then C operation to the decay: $K_L \rightarrow \pi^- + e^+ + v_{e\,LH}$, the decay process will become $K_L \rightarrow \pi^+ + e^- + v_e\grave{}_{RH,} >$. This can be written as:**

$CP \mid K_L \rightarrow \pi^- + e^+ + v_{e\,LH} > = \mid K_L \rightarrow \pi^+ + e^- + v_e\grave{}_{RH} >$

Thus, CP operation on $\pi^- + e^+ + v_{e\,LH}$ gives $\pi^+ + e^- + v_e\grave{}_{RH}$.

If weak interactions/decays were CP invariant, the both the decays $\pi^- + e^+ + v_{e\,LH}$ and $\pi^+ + e^- + v_e\grave{}_{RH}$ would occur in equal amounts and therefore both positrons e^+ and electrons e^- would produce in equal amount but the **positron is found to be more prolific by 0.33%.**
(The decay to the positron comes from K^0 component of the K_L beam and the decay to the electron comes from $K^{0\grave{}}$ component of the K_L beam).

For the Kaons, CP violation is a **tiny** effect in relatively **common** decays.
The decay $K_L \rightarrow \pi^- + e^+ + v_e$ is just 0.33% more common than its CP mirror image $K_L \rightarrow \pi^+ + e^- + v_e\grave{}$.
However, 40.55% of all long-lived Kaons K_L decay by leptonic modes.
For the B mesons, CP violation tends to be a **large** effect in extremely **rare** decays.
For example, the decay: **B^0 (db$\grave{}$) $\rightarrow K^+$ (us$\grave{}$) + π^- (u$\grave{}$d)** is 13% more common than its CP mirror image: **$B^{0\grave{}}$ (d$\grave{}$b) $\rightarrow K^-$ (u$\grave{}$s) + π^+ (ud$\grave{}$).**
However, the branching ratio for the decay: **B^0 (db$\grave{}$) $\rightarrow K^+$ (us$\grave{}$) + π^- (u$\grave{}$d)** is only 1.82 $\times 10^{-5}$.

note: What we define as matter or antimatter is just arbitrary. However K_L decay gives an unambiguous way of defining what we call matter or antimatter. **The positron (which is antimatter) is defined as the lepton which is more prolific in K_L decay.** So if an alien residing far away from the Earth somehow told us that electron is more prolific in neutral K_L decay in their planet, it means what we call as matter is assumed antimatter by them and therefore, if they say, they are made up of matter, it means, as per our definition, they are in fact made up of antimatter.

note: TCP or CPT theorem implies that all the interactions in nature are unchanged (invariant), if they are subjected to the combined operation of parity P (signs of the projections of spins of particles are changed), charge conjugation C (particles are replaced by antiparticles and vice versa) and reversal of time T.

K_L - K_S regeneration

If an initially pure neutral kaon K^0 beam **traverses in vacuum** for longer than the mean lifetime of short-lived Kaon K_S, then short-lived Kaon K_S disappears leaving a beam of pure long-lived Kaon K_L.

When the long-lived component K_L of a neutral kaon K^0 beam **traverses a slab of material,** then the components K^0 and $K^{0`}$ of $K_L$ are absorbed differently. The component $K^{0`}$ can also decay into the lambda baryon and positive pion after colliding with a proton, whereas the component K^0 cannot. Due to this, when K_L **comes out of the slab** from the other side, K_S also emerges that is, short-lived component K_S is regenerated.

(K_S was discovered in 1951 and K_L was discovered in 1956. K_L - K_S regeneration was observed in 1961).

note: Neutral Particle Oscillation is the transformation of a neutral particle into another neutral particle. There are two types of such oscillations.

1. Particle-antiparticle oscillation, e.g. **K^0 - K^0`** (neutral kaon – antineutral kaon) **Oscillation, B$^0$ - B$^0$`** (neutral B meson – antineutral B meson) **oscillation**.

2. Flavour oscillation, e.g. **neutrino oscillation.**

The particle-antiparticle oscillations proceed by a second order weak process that is, a particle is required to emit two W bosons to convert into its antiparticle. Quark flavour is not conserved in weak interactions, thus the transition between a neutral particle and its antineutral particle is possible via weak interactions.

D Mesons

The D mesons are the mesons containing charm quarks (charge: + 2/3e) or anti-charm quarks (charge: - 2/3e)

The combinations of the light quarks (d, u, s) and the anticharm quark and the combinations of the light antiquarks (d`, u`, s`) and charm quark form the charmed pseudoscalar D mesons and charmed vector D mesons. These combinations were observed close on the heels of the discovery of charmonium in 1974.

The pseudoscalar D mesons: positive D meson D^+ (cd`), negative D meson $D^-$ (c`d), neutral D meson D^0 (cu`), antineutral D meson $D^{0`}$ (c`u), positive strange D meson $D_s^+$ (cs`), negative strange D meson D_s^- (c`s).

To memorize D mesons, i.e. D^+ (cd`),  $D^-$ (c`d), D^0 (cu`), $D^{0`}$ (c`u), $D_s^+$ (cs`), D_s^- (c`s), you may do this.

Write d` d  u` u s`  s on a paper at a gap of some cms. Then. write c and c`alternately on the left side of each of these six. Close the brackets. Then write the symbol of corresponding D meson name on the left side of the brackets.

Note:

1. neutral kaon K^0 (ds`) and antineutral kaon $K^{0`}$ (d`s)
First symbol inside the bracket is d and d` respectively

2. neutral B meson B^0 (db`) and an antineutral B meson $B^{0`}$ (d`b)
First symbol inside the bracket is d and d` respectively

3. neutral D meson D^0 (cu`) and an antineutral D meson $D^{0`}$ (c`u)
First symbol inside the bracket is c and c` respectively

antineutral D meson $D^{0`}$ (c`u), has anti-charm c`. antineutral kaon $K^{0`}$ (d`s), has strange s.

The vector D mesons: positive D-star meson D*⁺ (cd`), negative D-star meson D*⁻ (c`d), neutral D-star meson D*⁰ (cu`), antineutral D-star meson D*⁰` (c`u), positive strange D-star meson D*ₛ⁺ (cs`), negative strange D-star meson D*ₛ⁻ (c`s).

Each of the pseudoscalar D mesons D⁺ (cd`) and D⁻ (c`d) has **mass** 1869 MeV and mean lifetime $\tau = 1.0 \times 10^{-12}$ s.

Each of the pseudoscalar D mesons D⁰ (cu`) and D⁰` (c`u) has **mass** 1864 MeV and mean lifetime $\tau = 0.4 \times 10^{-12}$ s.

Each of the pseudoscalar D mesons Dₛ⁺ (cs`) and Dₛ⁻ (c`s) has **mass** 1968 MeV and mean lifetime $\tau = 0.5 \times 10^{-12}$ s.

Each of the vector D mesons has rest mass of about 2000 MeV.

note: The positive as well as neutral D meson contains a charm quark c. The negative as well as antineutral D meson contains an anticharm quark c`.

The combination of a charm quark and an anticharm quark is not a neutral D meson, but rather **charmonium (cc`).**

Production of the positive D meson

When hydrogen bubble chamber BEBC (Big European Bubble Chamber) filled with superheated liquid hydrogen was exposed to a muon-type neutrino beam vµ at the CERN SPS, the collision between the muon-type neutrino vµ and the proton (of hydrogen atom) produced positive D meson:

$v_\mu + p \rightarrow \mu^- + D^+ (cd`) + p$

This is a weak interaction as charmness is not conserved. Charmness is C = 0 and +1 before and after the interaction respectively.

The muon-type neutrino v_μ emitted a W⁺ boson and transformed into a **muon µ⁻.** The W⁺ boson then decayed into a **charm quark c** (charge:

+2/3e) and an **antidown quark d`** (charge: +1/3e) which combined to produce a **positive D meson D⁺ (cd`).**

note: BEBC had been in use during 1973 to 1984. Charged particles passing through it would leave trails of bubbles.

Decay of the D Mesons

In a weak process, the **pseudoscalar D mesons preferably decay into the kaons** through W bosons with mean lifetime 10^{-12} s because D mesons contain charm quarks (charge: + 2/3e) or anti-charm quarks (charge: - 2/3e) and kaons contain strange quarks (charge: - 1/3e) or anti-strange quarks (charge: + 1/3e) and the charm and strange quarks are the same (second) generation quarks.

$|V_{cs}|^2$ implies the probability with which a charm quark c (charge: +2/3e) can transform into a strange quark s (charge: -1/3e), by emitting a W^+ boson or by absorbing a W^- boson **(c → s).**

$|V_{cs}|^2$ also implies the probability with which an anticharm quark c` (charge: -2/3e) transforms into an antistrange quark s` (charge: +1/3e) by emitting a W^- boson or by absorbing a W^+ boson **(c` → s`).**

$|V_{cs}|^2 = 0.986^2 = 0.972$ which is nearly equal to one.

(See the second note in the topic: Discovery of W and Z bosons).

D Mesons may decay as follow:

D⁺ (cd`) → D⁰ (cu`) + π⁺ (ud`); D⁺ (cd`) → K⁰` (d`s) + π⁺ (ud`) c → s through W boson

D⁰ (cu`) → K⁻ (u`s) + π⁺ (ud`) c → s through W boson;

D⁰` (c`u) → K⁺ (us`) + π⁻ (u`d) c`→ s` through W boson

D⁰ (cu`) → K⁺ (us`) + π⁻ (u`d) is CKM suppressed. c → d through W boson;

The positive D meson D⁺ (cd̀) subsequently decays into a neutral D meson D⁰ (cù) and a positive pion π⁺ (ud̀):

D⁺ (cd̀) → D⁰ (cù) + π⁺ (ud̀)

This is a strong decay (i.e. through gluon) as charmness is conserved.

In this case, the charm quark (charge: +2/3e) of the positive D meson emits a gluon and remains itself as a **charm quark c.**

The **gluon then decays into an up quark u and an antiup quark ù.**

This **antiup quark ù** combines with that **charm quark c** of the positive D meson which had emitted the gluon and produces a **neutral D meson D⁰ (cù)**.

The **up quark u** (produced by the decay of the gluon) combines with the **antidown quark d̀** of the positive D meson and produces a **positive pion π⁺ (ud̀)**.

The positive D meson D⁺ (cd̀) may <u>also decay</u> into an antineutral kaon K⁰̀ (d̀s) and a positive pion π⁺ (ud̀):

D⁺ (cd̀) → K⁰̀ (d̀s) + π⁺ (ud̀) c → s through W boson

This is weak decay (i.e. through W boson) as charmness and strangers are not conserved.

In this case, the charm quark (charge: +2/3e) of the positive D meson emits a **W⁺ boson** and **transforms into** a **strange quark s** (charge : - 1/3e) and the positive D meson transforms into an **antineutral kaon K⁰̀ (d̀s).**

The W⁺ boson then decays into an **up quark u** (charge: +2/3e) and an **antidown quark d̀** (charge: +1/3e) which combine to produce a **positive pion π⁺ (ud̀).**

The neutral D meson D^0 ($c\bar{u}$) produced by the decay of positive D meson D^+ ($c\bar{d}$) decays into a negative kaon K^- ($u\bar{s}$) and a positive pion π^+ ($u\bar{d}$):

D^0 ($c\bar{u}$) $\rightarrow$ K^- ($u\bar{s}$) + π^+ ($u\bar{d}$)   c $\rightarrow$ s through W boson

This is a weak decay (i.e. through W boson) as charmness and strangers are not conserved.

In this case, the **charm quark** (charge: +2/3e) of the neutral D meson **emits a W⁺ boson** and **transforms into** a **strange quark s** (charge : -1/3e) and the neutral D meson transforms into a **negative kaon K⁻** ($u\bar{s}$).

The W⁺ boson then decays into an **up quark u** (charge: +2/3e) and an **antidown quark d̀** (charge: +1/3e) which combine to produce a **positive pion π⁺** ($u\bar{d}$).

note:

5. The decay of the neutral D meson D^0 ($c\bar{u}$) into a positive kaon K^+ ($u\bar{s}$) and a negative pion π^- ($u\bar{d}$): D^0 ($c\bar{u}$) $\rightarrow$ K^+ ($u\bar{s}$) + π^- ($u\bar{d}$) is doubly CKM suppressed. (c $\rightarrow$ d through W boson and W⁺ $\rightarrow$ us̀)

This decay may be explained as:

The charm quark (charge: +2/3e) of the neutral D meson **emits a W⁺ boson** and **transforms into** a **down quark** (charge : -1/3e) and the neutral D meson transforms into a **negative pion π⁻** ($u\bar{d}$).

The W⁺ boson then decays into an **up quark u** (charge: +2/3e) and an **antistrange quark s̀** (charge: +1/3e) which combine to produce a **positive kaon K⁺** ($u\bar{s}$).

Now, note that the charm quark is a **second** generation quark and the down quark is a **first** generation quark. Thus, the **transformation of** this charm quark into a down quark **(c $\rightarrow$ d)** is **CKM suppressed.** ($|V_{cd}|^2 = 0.225^2 = 0.050$)

Similarly, the up quark is a **first** generation quark and the antistrange quark is a **second** generation antiquark. Thus, the **decay** of the W^+ boson into an up quark and an antistrange quark **($W^+ \rightarrow$ us`)** is also **CKM suppressed.** $(|V_{us}|^2 = 0.225^2 = 0.050)$

The decay of the **neutral D meson D^0 (cu`)** into a positive kaon $K^+$ (us`) and a negative pion π^- (u`d): **$D^0$ (cu`) $\rightarrow$ K^+ (us`) + $\pi^-$ (u`d) is doubly CKM suppressed. (c $\rightarrow$ d through W boson and $W^+ \rightarrow$ us`)**

<u>**Thus, the decay: D^0 (cu`) $\rightarrow$ $K^+$ (us`) + π^- (u`d) is doubly CKM suppressed. (c $\rightarrow$ d and $W^+ \rightarrow$ us`).**</u>

However, the decay $D^{0\text{`}}$ (c`u) $\rightarrow$ K^+ (us`) + $\pi^-$ (u`d) as well as the decay: D^0 (cu`) $\rightarrow$ $K^-$ (u`s) + π^+ (ud`) are not suppressed.

For example, In the decay: D^0 (cu`) $\rightarrow$ $K^-$ (u`s) + π^+ (ud`), the charm quark (second generation quark) transforms into a strange quark (second generation quark), for which the probability $|V_{cs}|^2 = 0.986^2 = 0.972$. Also, the W^+ boson decays into an up quark (first generation quark) and an antidown quark (first generation antiquark), for which the probability $|V_{ud}|^2 = 0.974^2 = 0.948$.

Elements of CKM matrix:

$V_{ud} = 0.974$ $V_{us} = 0.225$ $V_{ub} = 0.004$
$V_{cd} = 0.225$ $V_{cs} = 0.986$ $V_{cb} = 0.041$
$V_{td} = 0.008$ $V_{ts} = 0.040$ $V_{tb} = 0.999$

B Mesons

The B mesons are the mesons containing antibottom quarks (charge: + 1/3e)

The combinations of light quarks (u, d, s, c) and antibottom quark form the pseudoscalar B mesons and the vector B mesons.

The pseudoscalar B mesons: positive B meson B^+ (ub`), neutral B meson $B^0$ (db`), neutral strange B meson B_s^0 (sb`), positive charmed B meson $B_c^+$ (cb`)

To memorize B mesons, i.e. B^+ (ub`), $B^0$ (db`), B_s^0 (sb`), $B_c^+$ (cb`), you may do this.

Write u d   s c on a paper at a gap of some cms. Then. write b`in front of each of these four. Close the brackets. Then write the symbol of corresponding B meson name on the left side of the brackets.

note: antineutral D meson $D^{0`}$ (c`u), has anti-charm c`. antineutral kaon $K^{0`}$ (d`s), has strange s. antineutral B meson $B^{0`}$ (d`b) has bottom b.

The vector B mesons: positive B-star meson B^{*+} (ub`), neutral B-star meson $B^{*0}$ (db`), neutral strange B-star meson $B^*{}_s^0$ (sb`), positive charmed B-star meson $B^*{}_c^+$ (cb`)

The anti B mesons are the mesons containing bottom quarks (charge: - 1/3e)

The combinations of light antiquarks (u`, d`, s`, c`) and bottom quark form anti B mesons: negative B meson B^- (u`b), antineutral B meson $B^{0`}$ (d`b), antineutral strange B meson $B_s^{0`}$ (s`b), negative charmed B meson $B_c^-$ (c`b)

To memorize anti B mesons, i.e. B⁻ (u`b), B⁰` (d`b), B$_s$⁰` (s`b), B$_c$⁻ (c`b), you may do this.

Write u` d` s` c` on a paper at a gap of some cms. Then. write b in front of each of these four. Close the brackets. Then write the symbol of corresponding anti B meson name on the left side of the brackets.

Each of the pseudoscalar B mesons B⁺ (ub`) and B⁻ (u`b) has **mass** 5279 MeV and mean lifetime $\tau = 1.6 \times 10^{-12}$ s.

Each of the pseudoscalar B mesons B⁰ (db`) and B⁰` (d`b), has **mass** 5279 MeV and mean lifetime $\tau = 1.5 \times 10^{-12}$ s.

Each of the pseudoscalar B mesons B$_s$⁰ (sb`) and B$_s$⁰` (s`b) has **mass** 5366 MeV and mean lifetime $\tau = 1.5 \times 10^{-12}$ s.

Each of the pseudoscalar B mesons B$_c$⁺ (cb`) and B$_c$⁻ (c`b) has **mass** 6275 MeV and mean lifetime $\tau = 0.4 \times 10^{-12}$ s.

The combination of a bottom quark and an antibottom quark is not a neutral B meson, but rather **bottomonium (bb`).**

Decay of the B Mesons

In the decay of the B mesons, the antibottom quark b` (charge: +1/3e) transforms into an anticharm quark c` (charge: -2/3e) by emitting a W⁺ boson or by absorbing W⁻ boson **(b`    ›  c`).**

Third generation antibottom quark cannot decay into third generation antitop quark because this decay is kinematically forbidden as the mass of bottom and top quarks are 4200 MeV and 173 GeV respectively

In a weak process, the **pseudoscalar B mesons and anti B mesons predominantly decay into D (charmed) mesons** through W bosons with mean lifetime 10^{-12} s. In each of these decays, the third generation antibottom quark / bottom quark transforms **into** a second generation (intermediary mass) **anticharm quark / charm quark which is less CKM suppressed.**

D mesons contain charm quarks (charge: + 2/3e) or anti-charm quarks (charge: - 2/3e).

The pseudoscalar B mesons are B⁺ (ub`), B⁰ (db`), B$_s$⁰ (sb`), B$_c$⁺ (cb`)

The pseudoscalar B mesons may decay as follow:

B⁺ (ub`) → D⁰` (c`u) + π⁺ (ud`)   b` → c` through W boson

B⁰ (db`) → D⁻ (c`d) + π⁺ (ud`)    b` → c` through W boson

B$_s$⁰ (sb`) → D$_s$⁻ (c`s) + π⁺ (ud`)    b` → c` through W boson

B$_c$⁺ (cb`) → J/ψ (cc`) + π⁺ (ud`)    b` → c` through W boson

B⁰ (db`) → K⁺ (us`) + π⁻ (u`d)    b` → u` through W boson (CKM suppressed)

The positive B meson B⁺ (ub`) decays into an antineutral D meson D⁰` (c`u) and a positive pion π⁺ (ud`):

B⁺ (ub`) → D⁰` (c`u) + π⁺ (ud`) b` → c` through W boson

This is a weak decay (i.e. through W boson) as bottomness and charmness are not conserved.

Here, the **antibottom quark b`** (charge: +1/3e) of the positive B meson **emits a W⁺ boson and transforms into an anticharm quark c`**(charge: -2/3e). Thus, the positive B meson transforms into an **antineutral D meson D⁰` (c`u).**

The W⁺ boson then decays into an **up quark u** (charge: +2/3e) and an **antidown quark d`** (charge: -1/3e) which combine to produce a **positive pion π⁺ (ud`).**

The **neutral B meson B⁰ (db`)** decays into a negative D meson D⁻ (c`d) and a positive pion π⁺ (ud`):

B⁰ (db`) → D⁻ (c`d) + π⁺ (ud`) b` → c` through W boson

This is a weak decay (i.e. through W boson) as bottomness and charmness are not conserved.

The **neutral strange B meson B$_s^0$ (sb`)** decays into a negative strange D meson D$_s^-$ (c`s), and a positive pion π⁺ (ub`): **B$_s^0$ (sb`) → D$_s^-$ (c`s) + π⁺ (ud`) b` → c` through W boson**

This is a weak decay (i.e. through W boson) as bottomness and charmness are not conserved.

The **positive charmed B meson B$_c^+$ (cb`)** decays into a charmonium (cc`), and a positive pion π⁺ (ud`):

B$_c^+$ (cb`) → J/ψ (cc`) + π⁺ (ud`) b` → c` through W boson

This is a weak decay (i.e. through W boson) as bottomness and charmness are not conserved.

(Mass of B$_c^+$ (6276 MeV) > mass of J/ψ (3.1 GeV) + mass of positive pion (139 MeV), thus this decay is kinematically allowed).

note:
The positive B meson B⁺ (ub`) is similar to the positive pion π⁺ (ud`) except that the antidown quark d` is replaced by the antibottom quark b`.

note:
The decay of the **neutral D meson D⁰ (cu`)** into a positive kaon K⁺ (us`) and a negative pion π⁻ (u`d): **D⁰ (cu`) → K⁺ (us`) + π⁻ (u`d) is doubly CKM suppressed. (c → d through W boson and W⁺ → us`)**

Similarly, the decay of the **neutral B meson B⁰ (db`)** into a positive kaon K⁺ (us`) and a negative pion π⁻ (u`d): **B⁰ (db`)** → K⁺ (us`) + π⁻ (u`d) is doubly CKM suppressed. (b` → u` through W boson and W⁺ → us`)

The branching ratio for the decay: B⁰ (db`) → K⁺ (us`) + π⁻ (u`d) is only 1.82 ×10⁻⁵.

This decay may be explained as:

This is a weak decay (i.e. through W boson) as bottomness and strangeness are not conserved.

The strange quark has strangeness quantum number **S = -1,** The bottom quark has bottomness quantum number **B` = -1,**

Strangeness is S = 0 and +1 before and after the decay respectively,

Bottomness is B` = +1 and 0 before and after the decay respectively.

The **antibottom quark b`** (charge: +1/3e) of the neutral B meson **emits a W⁺ boson** (and therefore emits charge: +1e) and **transforms into an antiup quark u`** (charge +1/3e -1e = -2/3e) and the neutral B meson transforms into a **negative pion π⁻ (u`d).**

The W⁺ boson subsequently decays into an **up quark u** (charge: +2/3e) and an **antistrange quark s`** (charge: +1/3e) which combine to produce a **positive kaon K⁺ (us`).**

Now, note that the antibottom quark is a **third** generation antiquark and the antiup quark is a **first** generation antiquark. Thus, the **transformation** of this antibottom quark into antiup quark (b` → u`) is **highly CKM suppressed.** ($|V_{ub}|^2 = 0.004^2 = 0.000016$).

In above, actually ($|V_{ub}|^2$ = is for u` → b` or u → b

Similarly, the up quark is a **first** generation quark and the antistrange quark is a **second** generation antiquark. Thus, the **decay** of the W⁺ boson into an up quark and an antistrange quark (**W⁺ → us`**) is also **CKM suppressed.** ($|V_{us}|^2 = 0.225^2 = 0.050$)

Thus, the decay: **B⁰ (db`) → K⁺ (us`) + π⁻ (u`d) is doubly CKM**

suppressed. (b` → u` and W⁺ → us`).

(See the notes at the end of the topic: Discovery of W and Z bosons)

B⁰ - B⁰` (neutral B meson – antineutral B meson) **Mixing**

The neutral B meson B⁰ (db`) can transform into the antineutral B

meson B⁰` (d`b) while propagating in space and vice versa.

In this transformation, bottomness is not conserved as the bottomness

of the neutral B meson B⁰ (db`) is B` = +1 and the bottomness of the

antineutral B meson B⁰` (d`b) is B` = -1, thus this transformation is

possible through the weak process.

The **down quark d (charge: -1/3e)** of the neutral B meson **emits a W⁻**

boson and transforms into a top quark t (charge: +2/3e). This top quark

t (charge: +2/3e) and the antibottom quark b` (charge: +1/3e) of the

neutral B meson **annihilate to a W⁺ boson**.

Thus, in this way, the down quark d and the antibottom quark b`of the

neutral B meson B⁰ (db`) **radiate two W bosons** and **exchange a top**

quark between them.

The **W⁺ boson decays** into an **antidown quark d`** (charge: +1/3e) and

a top quark t (charge: +2/3e). This top quark t (charge: +2/3e) **absorbs**

W⁻ boson and transforms into a **bottom quark b** (charge: -1/3e). Thus,

the two W bosons **exchange a top quark** between them.

The **antidown quark d`** and the **bottom quark b** combine to produce

the **antineutral B meson B⁰` (d`b).**

OR

The **antibottom quark b` (charge: +1/3e)** of the neutral B meson

emits a W⁺ boson and transforms into an antitop quark t` (charge: -

2/3e). This antitop quark t` (charge: -2/3e) and down quark d (charge: -

1/3e) of the neutral B meson **annihilate to a W⁻ boson.** Thus, in this

way too, the down quark d and the antibottom quark b`of the neutral B meson $B^0$ (db`) may **radiate two W bosons** and **exchange an antitop quark** between them.

The **W⁻ boson decays** into a **bottom quark b** (charge: -1/3e) and an antitop quark t` (charge: -2/3e). This antitop quark t` (charge: -2/3e) **absorbs W⁺ boson** and transforms into an **antidown quark d** (charge: +1/3e). Thus, the two W bosons **exchange an antitop quark** between them.

The **antidown quark d`** and the **bottom quark b** combine to produce the **antineutral B meson $B^{0`}$ (d`b).**

Similarly, the antineutral B meson $B^{0`}$ (d`b) can transform into the neutral B meson B^0 (db`) as follows:

The **bottom quark b (charge: -1/3e)** of the antineutral B meson **emits a W⁻ boson** and transforms into a top quark t (charge: +2/3e). This top quark t (charge: +2/3e) and the antidown quark d` (charge: +1/3e) of the antineutral B meson **annihilate to a W⁺ boson.**

Thus, in this way, the antidown quark d` and the bottom quark b of the antineutral B meson $B^{0`}$ (d`b) **radiate two W bosons** and **exchange a top quark** between them.

The **W⁺ boson decays** into an **antibottom quark b`** (charge: +1/3e) and a top quark t (charge: +2/3e). The top quark t (charge: +2/3e) **absorbs W⁻ boson** and transforms into a **down quark d** (charge: -1/3e). Thus, the two W bosons **exchange a top quark** between them.

The **down quark d** and the **antibottom quark b`** combine to produce the **neutral B meson $B^0$ (db`)**

B^0 - $B^{0`}$ **Mixing is dominated by top quark (or antitop quark) exchange** as explained above. The down quark and the antibottom quark of the neutral B meson $B^0$ (db`) may exchange up or charm quark too and similarly the antidown quark and the bottom quark of the antineutral B meson **B⁰`** (d`b) may exchange up or charm quark too but mixing through those exchanges is minute.

B$_s^0$ - B$_s^{0`}$ (neutral strange B meson - antineutral strange B meson)
Mixing: The neutral strange B meson B$_s^0$ (sb`) can transform into the antineutral strange B meson B$_s^{0`}$ (s`b) while propagating in space and vice versa.
This mixing can take place in the same way as explained for **B⁰ - B⁰`**
Mixing.

note: D⁰ - D⁰` (neutral D meson - antineutral D meson) **Mixing** is minute and is unobservable.

Unlike the decay, K⁺ (us`) → μ⁺ + ν$_μ$, **the decay B⁺ (ub`) → μ⁺ + ν$_μ$ is highly CKM suppressed:**
The up quark is the **first** generation quark and the antibottom quark is the **third** generation antiquark. ($|V_{ub}|^2$ = 0.004² = 0.000016).
Thus, the **annihilation** of the up quark u (charge: +2/3e) and the antibottom quark b` (charge: +1/3e) of the positive B meson B⁺ (ub`) to a W⁺ boson **(ub` → W⁺)** is **highly CKM suppressed**
Thus, the positive B meson B⁺ (ub`) cannot decay into μ⁺ and ν$_μ$ through the W⁺ boson because the W⁺ boson cannot act as a mediator for this decay in the first place.
(See the notes at the end of the topic: Discovery of W and Z boson)

The decay B⁰ (db`) → μ⁺ + μ⁻ is highly suppressed:
(The branching ratio for the decay: B⁰ (db`) → μ⁺μ⁻ is 1.06 × 10⁻¹⁰)
The Z⁰ boson cannot couple directly to the quarks of different flavours, i.e. no direct 'flavour changing neutral current' exists.

Thus, the down quark d (charge: -1/3e) and the antibottom quark b`
(charge: +1/3e) of the neutral B meson B^0 (db`) **cannot annihilate** to a
Z^0 boson (**db`** → **Z^0**).

Thus, the neutral B meson B^0 (db`) cannot decay into $\mu^+\mu^-$ through the
Z^0 boson as the Z^0 boson cannot act as a mediator for this decay in the
first place.

Moreover, the down quark and the antibottom quarks are **first**
generation quark and a **third** generation antiquark respectively. **Thus,
this decay is CKM suppressed too.**

The decay B_s^0 (sb`) → $\mu^+ + \mu^-$ is highly suppressed:
(The branching ratio for the decay: B_s^0 (sb`) → $\mu^+\mu^-$ is 3.66 × 10^{-9})

The Z^0 boson cannot couple directly to the quarks of different flavours,
i.e. no direct 'flavour changing neutral current' exists.

Thus, the strange quark s (charge: -1/3e) and the antibottom quark
b`(charge: +1/3e) of the neutral strange B meson $B_s^0$ (sb`) **cannot
annihilate** to a Z^0 boson (**sb`** → **Z^0**).

Thus, the neutral strange B meson B_s^0 (sb`) cannot decay into $\mu^+\mu^-$
through the Z^0 boson as the Z^0 boson cannot act as a mediator for this
decay in the first place.

Moreover, the strange quark and the antibottom quarks are **second**
generation quark and a **third** generation antiquark respectively. **Thus,
this decay is CKM suppressed too.**

**The decay of the neutral B meson B^0 (db`) and the neutral strange
B meson B_s^0 (sb`) into muon-antimuon pairs are extremely rare.**
Extremely rarely, the neutral strange B meson B_s^0 (sb`) **may decay
into muon-antimuon pair $\mu^+\mu^-$ through higher order weak process**
as follows

The **strange quark s (charge: -1/3e)** of the neutral strange B meson **emits a W⁻ boson** and transforms into a top quark t (charge: +2/3e). This top quark t (charge: +2/3e) and the antibottom quark b˙ (charge: +1/3e) of the neutral strange B meson **annihilate to a W⁺ boson**. In this way, the strange quark s and the antibottom quark b˙ of the neutral strange B meson B_s^0 (sb˙) **radiate two W bosons** and **exchange a top quark** between them.

The **W⁺ boson decays** into an **antimuon μ⁺** and a muon-type neutrino v_μ. The muon-type neutrino v_μ may then absorb W⁻ boson and transform into a **muon μ⁻**. Thus, the two W bosons **exchange a muon-type neutrino v_μ** between them.

Instead of exchanging **muon-type neutrino v_μ**, W bosons may fuse to a **Z⁰ boson which** then decays into **muon-antimuon pair μ⁺μ⁻** in the final state.

note:

The decay B_s^0 (sb`) $\rightarrow \mu^+\mu^-$ through higher order weak process (explained above) is rare, as even this decay process is CKM suppressed.

Also, the decay of B_s^0 (sb`) through higher order weak process too is highly suppressed as this weak decay involves additional exchanges/annihilations of particles which reduces extremely the branching ratio or the probability of occurring the decay in this way.

The branching ratio for the decay: B_s^0 (sb`) $\rightarrow \mu^+\mu^-$ is 3.66×10^{-9}, i.e. 1 B_s^0 will decay into $\mu^+\mu^-$ for each $1/(3.66 \times 10^{-9})$ B_s^0 produced, i.e. **for every 1 billion (10^9) neutral strange B meson B_s^0 produced, 3 decay into $\mu^+\mu^-$.**

The decay B^0 (db`) $\rightarrow \mu^+\mu^-$ through higher order weak process is even more CKM suppressed than the decay $B_s^0$ (sb`) $\rightarrow \mu^+\mu^-$ as B^0 (db`)** consists of first generation quark and third generation antiquark whereas **$B_s^0$ (sb`) consists of second generation quark and third generation antiquark. (**B^0 (db`)** requires a jump across two quark generations rather than just one).

Like the decay: B_s^0 (sb`) $\rightarrow \mu^+\mu^-$, the decay $B^0$ (db`) $\rightarrow \mu^+\mu^-$ through higher order weak process too is highly suppressed as this weak decay involves additional exchanges/annihilations of particles which reduces extremely the branching ratio or the probability of occurring the decay in this way.

The branching ratio for the decay: B^0 (db`) $\rightarrow \mu^+\mu^-$ is 1.06×10^{-10}, i.e. 1 B^0 will decay into $\mu^+\mu^-$ for each $1/(1.06 \times 10^{-10})$ B_s^0 produced, i.e. **for every 10 billion (10^{10}) neutral B meson B^0 produced, 1 decays into $\mu^+\mu^-$.**

More than 90% of **B$_s^0$** meson decay into a D meson and other particles, e.g. B$_s^0$ (sb`) → D$_s^-$ (c`s) + π$^+$ (ud`) but a tiny fraction of B$_s^0$ mesons decay into muon-antimuon pairs.

note:

The symbols N, Δ, Λ, Σ, Ξ, and Ω are used for the baryons made up of light quarks (u, d and s quarks).

Baryons with three up and/or down quarks are nucleons or delta baryons.

Baryons with two up and/or down quarks are sigma baryons or lambda baryons. The third quark in them is strange quark. If the third quark is heavy (c or b), a subscript is used to identify it, e.g. Σ$_c^{++}$ (uuc), Λ$_c^+$ (udc).

Baryons with one up and/or down quark are Xi baryons. The remaining two quarks in them are strange quarks. If one or both of the remaining quarks are heavy (c or b), one or two subscript are used to identify it, e.g. Ξ$_c^0$ (dsc), Ξ$_{cc}^+$ (dcc).

Baryons with no up or down quark are omega baryons. Negative omega baryon has 3 strange quarks. If one or two or three quarks are heavy (c or b), one or two or three subscripts are used to identify it, e.g. Ω$_{cc}^+$ (scc), Ω$_{ccc}^{++}$ (ccc).

Lambda Baryon Λ^0 (uds) **consists of** one up quark, one down quark and **one strange quark.**

Charmed Lambda means strange quark replaced by charm quark. Thus, **Charmed Lambda Baryon** Λ$_c^+$ (udc) consists of one up quark, one down quark and **one charm quark.**

Bottom Lambda means strange quark replaced by bottom quark. Thus, **Bottom Lambda Baryon** Λ$_b^0$ (udb) consists of one up quark, one down quark and **one bottom quark.**

Each of the three **Sigma Baryons** Σ^+ (uus), Σ^0 (uds), Σ^- (dds) **consists of** two lightest quarks (up and/or down) and one strange quark.

Charmed Sigma means strange quark replaced by charm quark.

Thus, **each** of the three **Charmed Sigma Baryons** Σ_c^{++} (uuc), Σ_c^+ (udc), Σ_c^0 (ddc) consists of two lightest quarks (up and/or down) and **one charm quark.**

Bottom Sigma means strange quark replaced by bottom quark.

Thus, **each** of three **Bottom Sigma Baryons** Σ_b^+ (uub), Σ_b^0 (udb), Σ_b^- (ddb) consists of two lightest quarks (up and/or down) and **one bottom quark.**

Each of the two **Xi Baryons** Ξ^0 (uss), Ξ^- (dss) **consists of** one lightest quark (up and/or down) and two strange quarks.

Charmed Xi means one strange quark replaced by one charm quark.

Thus, **each** of the two **Charmed Xi Baryons** Ξ_c^+ (usc), Ξ_c^0 (dsc) consists of one lightest quark (up and/or down). **one strange quark and one charm quark**.

Doubled Charmed Xi means two strange quarks replaced by two charm quarks.

Thus, **each** of the two **Doubled Charmed Xi Baryons** Ξ_{cc}^{++} (ucc), Ξ_{cc}^+ (dcc) consists of one lightest quark (up and/or down) and **two charm quarks.**

Bottom Xi means one strange quark replaced by one bottom quark.

Each of the two **Bottom Xi Baryons:** Ξ_b^0 (usb), Ξ_b^- (dsb) consists of one lightest quark (up and/or down), **one strange quark and one bottom quark.**

Supermultiplets

Baryon 20-plet and Baryon 20′ -plet

Just as the combinations of u, d, s quarks form baryon octet (also called SU(3) octet) and baryon decuplet (also called SU(3) decuplet), the combinations of u, d, s, c quarks form **baryon 20-plet** (also called SU(4) 20-plet) and **baryon 20′ -plet** (also called SU(4) 20′ -plet).

In 20′ -plet (20 prime plet), there are 20 baryons, each with spin 1/2. These 20 (8 + 9 + 3) baryons are divided into 3 groups:

In the first group, there are **eight baryons of baryon octet.** (See the topic: The Baryon Octet):

n (udd), p (uud)

Σ^- (dds), Σ^0 (uds), Λ^0 (uds), Σ^+ (uus)

Ξ^- (dss), Ξ^0 (uss)

In the second group, there are **nine baryons (each having one charm quark):**

Σ_c^0 (ddc), Σ_c^+ (udc), Λ_c^+ (udc), Σ_c^{++} (uuc)   (written from the second line of the baryon octet by replacing s by c)

$\Xi'_c{}^0$ (dsc), Ξ_c^0 (dsc), $\Xi'_c{}^+$ (usc), Ξ_c^+ (usc)   (written from the third line of the baryon octet by replacing s by c)

Ω_c^0 (ssc)

That is, the second group include following nine baryons:

Three charmed sigma baryons: double positive charmed sigma baryon Σ_c^{++} **(uuc)**, positive charmed sigma baryon Σ_c^+ **(udc)**, neutral charmed sigma baryon Σ_c^0 **(ddc).**

The composition of these three charmed sigma baryons is similar to that of three sigma baryons of baryon octet except that one s is replaced by one c here.

One charmed lambda baryon: positive charmed lambda baryon Λ_c^+ **(udc).**

The composition of charmed lambda baryons is similar to that of lambda baryon of baryon octet except that s is replaced by c here.

Four charmed Xi baryons: positive charmed Xi baryon Ξ_c^+ **(usc),** positive charmed prime Xi baryon Ξ'_c^+ **(usc),** neutral charmed Xi baryon Ξ_c^0 **(dsc),** neutral charmed prime Xi baryon Ξ'_c^0 **(dsc).**

The composition of these four charmed Xi baryons is similar to that of two Xi baryons of baryon octet except that one s is replaced by one c here.

One charmed omega baryon: neutral charmed omega baron Ω_c^0 **(ssc).**

In the third group, there are **three baryons (each having two charm quark).**

Ξ_{cc}^+ **(dcc)** Ξ_{cc}^{++} **(ucc)** (written from the third line of the baryon octet by replacing ss by cc)

Ω_{cc}^+ **(scc)**

That is, the third group include following three baryons:

Two double charmed Xi baryons: double positive double charmed Xi baryon Ξ_{cc}^{++} **(ucc),** positive double charmed Xi baryon Ξ_{cc}^+ **(dcc).**

One double charmed omega baryon: positive double charmed omega baryon Ω_{cc}^+ **(scc).**

note:

Ξ'_c^+ (2575.6 MeV) is electromagnetic excited state of Ξ_c^+ (2467.8 MeV) and decays into Ξ_c^+: Ξ'_c^+ **(usc)** $\rightarrow$ Ξ_c^+ **(usc) + γ**

Ξ'^0_c (2577.9 MeV) is electromagnetic excited state of Ξ^0_c (2470.88 MeV) and decays into Ξ^0_c: $\mathbf{\Xi'^0_c\,(dsc)} \rightarrow \mathbf{\Xi^0_c\,(dsc) + \gamma}$

In 20-plet, there are 20 baryons, each with spin 3/2. These 20 (10 + 6 + 3 +1) baryons are divided into 4 groups:

In the first group, there are **ten baryons of baryon decuplet** (See the topic: The Baryon decuplet):

Δ^- (ddd), Δ^0 (udd), Δ^+ (uud), Δ^{++} (uuu)

Σ^{*-} (dds), Σ^{*0} (uds), Σ^{*+} (uus)

Ξ^{*-} (dss), Ξ^{*0} (uss)

Ω^- (sss)

In the second group, there are **six baryons (each having one charm quark):**

Σ_c^{*0} (ddc), Σ_c^{*+} (udc), Σ_c^{*++} (uuc)   (written from the second line of the baryon decuplet by replacing s by c)

Ξ_c^{*0} (dsc), Ξ_c^{*+} (usc)   (written from the third line of the baryon decuplet by replacing s by c)

Ω_c^{*0} (ssc)   (written from the fourth line of the baryon decuplet by replacing s by c)

That is, the second group include following six baryons:

Three charmed sigma-star baryons: double positive charmed sigma-star baryon Σ_c^{*++} (uuc), positive charmed sigma-star baryon Σ_c^{*+} (udc), neutral charmed sigma-star baryon Σ_c^{*0} (ddc).

The composition of these three charmed sigma baryons is similar to that of three sigma-star baryons of decuplet except that one s is replaced by one c here.

Two charmed Xi-star baryons: positive charmed Xi-star baryon Ξ_c^{*+} (usc), neutral charmed Xi-star baryon Ξ_c^{*0} (dsc).

The composition of these two charmed Xi-star baryons is similar to that of two Xi-star baryons of decuplet except that one s is replaced by one c here.

One charmed omega-star baryon: neutral charmed omega-star baryon Ω^{*0}_{c} (ssc).

The composition of this charmed omega-star baryon is similar to that of omega baryon of decuplet except that one s is replaced by one c here.

In the third group, there are **three baryons (each having two charm quark):**

Ξ^{*+}_{cc} (dcc), Ξ^{*++}_{cc} (ucc)   (written from the third line of the baryon decuplet by replacing ss by cc)

Ω^{*+}_{cc} (scc)   (written from the fourth line of the baryon decuplet by replacing ss by cc)

That is, the third group include following three baryons:

Two double charmed Xi-star baryons: double positive double charmed Xi-star baryon Ξ^{*++}_{cc} (ucc), positive double charmed Xi-star baryon Ξ^{*+}_{cc} (dcc)

The composition of these two double charmed Xi-star baryons is similar to that of two Xi-star baryons of decuplet except that ss are replaced by cc here.

One double charmed omega-star baryon: positive double charmed omega-star baryon Ω^{*+}_{cc} (scc)

The composition of this double charmed omega-star baryon is similar to that of omega baryon of decuplet except that ss is replaced by cc here.

In the fourth group, there ia **one baryon (having three charm quark):**

Ω^{++}_{ccc} (ccc)   (written from the fourth line of the baryon decuplet by replacing sss by ccc)

Triple charmed omega baryon: double positive triple charmed omega baryon Ω^{++}_{ccc} (ccc).

note:

The positive charmed Xi-star baryon $\Xi^*_c{}^+$ (usc) or or $\Xi_c(2645)^+$ **has rest mass** 2645.9 MeV and decays **into a neutral charmed Xi baryon $\Xi_c{}^0$ (dsc) and a positive pion π^+ (ud`):**

$$\Xi^*_c{}^+ (usc) \rightarrow \Xi_c{}^0 (dsc) + \pi^+ (ud`)$$

In this decay, the **up quark u** (charge: +2/3e) of the positive charmed Xi-star baryon **emits a gluon** and remains itself as an **up quark u.**

The **gluon then decays into a down quark d and an antidown quark d`.** This **antidown quark d`** combines with that **up quark u** of the positive charmed Xi-star baryon which had emitted the gluon and produces a **positive pion π^+ (ud`).**

The **down quark d** (produced by the decay of the gluon) combines with the **strange quark s** and the **charm quark c** of the positive charmed Xi-star baryon and produces **a neutral charmed Xi baryon $\Xi_c{}^0$ (dsc)**

The neutral charmed omega-star baryon $\Omega^*_c{}^0$ **(ssc) or** $\Omega(2770)^0$ decays into a neutral charmed omega baryon $\Omega_c{}^0$ **(ssc)** and a photon:

$$\Omega^*_c{}^0 \text{ (ssc)} \rightarrow \Omega_c{}^0 \text{ (ssc)} + \gamma$$

The Mass difference (2765.9 MeV - 2695.2 MeV = 70.7 MeV) of $\Omega^*_c{}^0$ (ssc) and $\Omega_c{}^0$ (ssc) is too small for any strong decay to occur.

Also, $\Omega^*_c{}^0$ (ssc) cannot decay into $\Xi_c{}^+$ (2467.8 MeV) and K^- (493.6) by emitting a gluon (which would decay into up and antiup quark), since 2765.9 MeV < 2467.8 + 493.6. Thus, the decay, $\Omega^*_c{}^0$ **(ssc)** $\rightarrow \Xi_c{}^+$ (usc) + K^- (u`s) is kinematically forbidden.

Meson 16-plet and Meson 16′ -plet.

Just as combinations of u, d, s quarks form two meson nonets, combination of u, d, s, c quarks form **meson 16-plet** and **meson 16′ - plet**.

In 16′ -plet (16 prime plet), there are 16 pseudoscalar mesons (spin 0). These 16 (10 + 3 + 3) pseudoscalar mesons are divided into 3 groups:

In the first group, there are **nine mesons of the pseudoscalar meson nonet** (See the topic: The Pseudoscalar Meson Nonet) and **charmed eta meson η_c (cc`).**

In the second group, there are **three D mesons i.e. positive D meson D^+ (cd`), neutral D meson $D^0$ (cu`), positive strange D meson D_s^+ (cs`).**

In the third group, there are other **three D mesons i.e. negative D meson D^- (c`d), antineutral D meson $D^{0`}$ (c`u), negative strange D meson $D_s^-$ (c`s).**

In 16-plet, there are 16 vector mesons (spin 1). These 16 (10 + 3 + 3) vector mesons are divided into 3 groups:

In the first group, there are **nine mesons of the vector meson nonet** (See the topic: The Vector Meson Nonet) and **J/ψ meson (cc`).**

In the second group, there are **three D-star mesons i.e. positive D-star meson D^{*+} (cd`), neutral D-star meson $D^{*0}$ (cu`), positive strange D-star meson $D^*_s{}^+$ (cs`).**

In the third group, there are other **three D-star mesons i.e. negative D-star meson D*⁻ (c`d), antineutral D-star meson D\*⁰` (c`u), negative strange D-star meson D\*ₛ⁻ (c`s).**

Charmed and Bottom Lambda Baryons

1. The decay of the positive charmed lambda baryon Λ_c^+ (udc) into a proton p (uud), a negative kaon K^- (u`s) and a positive pion $\pi^+$ (ud`):

Λ_c^+ (udc) $\rightarrow$ p (uud) + K^- (u`s) + $\pi^+$ (ud`)

In this decay, charmness and strangeness are not conserved. Thus, this is a weak decay i.r. through W boson.

In this decay, the **charm quark c** (charge: +2/3e) of the positive charmed lambda baryon **emits a W⁺ boson** and **transforms into** a **strange quark s** (charge: -1/3e).

The W⁺ boson then decays into an **up quark u** (charge: +2/3e) and an **antidown quark d`** (charge: +1/3e) which combine to produce a **positive pion $\pi^+$ (ud`).**

The **up quark** of the positive charmed lambda baryon **emits a gluon** and the up quark remains itself as an **up quark u.**

The **gluon then decays into an up quark u and an antiup quark u`.**

This **antiup quark u`** combines with the **strange quark s** (produced by the transformation of the charm quark) and produces a **negative kaon $K^-$ (u`s).**

The **up quark u** (produced by the decay of the gluon) combines with the **down quark d** of the positive charmed lambda baryon and that **up quark u** of the positive charmed lambda baryon which had emitted the gluon and produces a **proton p (uud).**

2. The decay of the neutral bottom lambda baryon Λ_b^0 (udb) into a lambda baryon Λ^0 (uds) and a J/ψ meson (cc`):

Λ_b^0 (udb) $\rightarrow$ Λ^0 (uds) + J/ψ (cc`)

In this decay, strangeness and bottomness are not conserved. Thus, it is a weak decay, i.e. through W boson.

In this decay, the **bottom quark b** (charge: -1/3e) of the neutral bottom lambda baryon **emits a W⁻ boson** (and therefore emits charge: -1e) and **transforms into** a **charm quark c** (charge: -1/3e +1e = +2/3e). The W⁻ boson then decays into a **strange quark s** (charge: -1/3e) and an **anticharm quark c`** (charge: -2/3e). This **anticharm quark c`** combines with the **charm quark c** (produced by the transformation of the bottom quark) and produces a **J/ψ meson (cc`)**.

The **strange quark s** (produced by the decay of the W⁻ boson) combines with the up quark and the down quark of the neutral bottom lambda baryon and produces a **lambda baryon Λ⁰ (uds)**.

note: The neutral bottom lambda baryon Λ_b^0 (udb) was observed in the 1980s. This was the first baryon observed which had a bottom quark.

3. The decay of the neutral bottom lambda baryon Λ_b^0 (udb) into a positive charmed lambda baryon Λ_c^+ (udc) and a negative pion π⁻ (u`d):

Λ_b^0 (udb) → Λ_c^+ (udc) + π⁻ (u`d)

In this decay, bottomness and charmness are not conserved. Thus, this is a weak decay, i.e. through W boson.

In this decay, the **bottom quark b** (charge: -1/3e) of the neutral bottom lambda baryon **emits a W⁻ boson** and **transforms into** a **charm quark c** (charge: +2/3e) and the neutral bottom lambda baryon transforms into a **positive charmed lambda baryon Λ_c^+ (udc)**.

The W⁻ boson then decays into an **antiup quark u`** (charge: -2/3e) and a **down quark d** (charge: -1/3e) which combine to produce a **negative pion π⁻ (u`d)**.

4. The decay of the neutral bottom lambda baryon Λ_b^0 (udb) into a positive charmed lambda baryon Λ_c^+ (udc) and a negative strange D meson D_s^- (c`s):

Λ_b^0 (udb) $\rightarrow$ Λ_c^+ (udc) + D_s^- (c`s)

In this decay, bottomness and charmness are not conserved.
Strangeness is S = 0 and -1 before and after the decay respectively.
Thus this decay is through the weak process that is, through a W
boson.

In this case, the **bottom quark b** (charge: -1/3e) of the neutral bottom
lambda baryon **emits a W⁻ boson** and **transforms into** a **charm quark
c** (charge: +2/3e) and the neutral bottom lambda baryon transforms into
a **positive charmed lambda baryon Λ_c^+ (udc).**

The W⁻ boson then decays into a **strange quark s** (charge: -1/3e) and
an **anticharm quark c`** (charge: -2/3e) which combine to produce a
negative strange D meson D_s^- (c`s).

Charmed and Bottom Sigma Baryons

1. The decay of the double positive charmed sigma baryon Σ_c^{++} (uuc) into a positive charmed lambda baryon Λ_c^+ (udc) and a positive pion π^+ (ud`):

Σ_c^{++} (uuc) $\rightarrow$ Λ_c^+ (udc) + π^+ (ud`)

In this case, charmness is conserved. Charmness is C = +1 before as well as after the decay. Thus, this decay is through the strong process, i.e. through gluon.

In this decay, one **up quark** of the double positive charmed sigma baryon **emits a gluon** and the up quark remains itself as an **up quark u**.

The **gluon then decays into a down quark d and an antidown quark d`**. This **antidown quark d`** combines with that **up quark u** of the double positive charmed sigma baryon which had emitted the gluon and produces a **positive pion π^+ (ud`).**

The **down quark d** (produced by the decay of the gluon) combines with the remaining up quark and the charm quark of the double positive charmed sigma baryon and produces a **positive charmed lambda baryon Λ_c^+ (udc).**

2. The decay of the positive charmed sigma baryon Σ_c^+ (udc) into a positive charmed lambda baryon Λ_c^+ (udc) and a neutral pion π^0 (uu`):

Σ_c^+ (udc) $\rightarrow$ Λ_c^+ (udc) + π^0 (uu`)

In this case too, charmness is conserved.

In this decay, the **up quark** of the positive charmed sigma baryon **emits a gluon** and the up quark remains itself as an **up quark u**.

The **gluon then decays into a up quark u and an antiup quark u`**.

This **antiup quark u`** combines with that **up quark u** of the positive charmed sigma baryon which had emitted the gluon and produces a **neutral pion $\pi^0$ (uu`)**.

The **up quark u** (produced by the decay of the gluon) combines with the down quark and the charm quark of the positive charmed sigma baryon and produces a **positive charmed lambda baryon Λ_c^+ (udc)**. Through the emission of a gluon, Σ_c^+ **(2452.9 MeV)** loses energy and transforms into Λ_c^+ **(2286.4 MeV) which** contains the same composition of the fundamental particles as Σ_c^+, i.e. one up quark, one down quark and one strange quark.

3. The decay of the neutral charmed sigma baryon Σ_c^0 (ddc) into a positive charmed lambda baryon Λ_c^+ (udc) and a negative pion π^- (u`d):

Σ_c^0 **(ddc)** $\rightarrow \Lambda_c^+$ **(udc)** + π^- **(u`d)**

In this case too, charmness is conserved.

In this decay, one **down quark** of the neutral charmed sigma baryon **emits a gluon** and the down quark remains itself as a **down quark d**. The **gluon then decays into an up quark u and an antiup quark u`**. This **antiup quark u`** combines with that **down quark d** of the neutral charmed sigma baryon which had emitted the gluon and produces a **negative pion π^- (u`d)**.

The **up quark u** (produced by the decay of the gluon) combines with the remaining down quark and the charm quark of the neutral charmed sigma baryon and produces a **positive charmed lambda baryon Λ_c^+ (udc)**.

4. The decay of the positive bottom sigma baryon Σ_b^+ (uub) into a neutral bottom lambda baryon Λ_b^0 (udb) and a positive pion π^+ (ud`):

Σ_b^+ **(uub)** $\rightarrow \Lambda_b^0$ **(udb)** + π^+ **(ud`)**

In this decay, bottomness is conserved. Thus, this is a strong decay, i.e. through gluon.

In this decay, one **up quark** of the positive bottom sigma-star baryon **emits a gluon** and the up quark remains itself as an **up quark u**.
The **gluon then decays into a down quark d and an antidown quark d`**. This **antidown quark d`** combines with that **up quark u** of the positive bottom sigma baryon which had emitted the gluon and produces a **positive pion π^+ (ud`)**.
The **down quark d** (produced by the decay of the gluon) combines with the remaining up quark and the bottom quark of the positive bottom sigma baryon and produces a **neutral bottom lambda baryon $\Lambda_b{}^0$ (udb)**.

5. The decay of the negative bottom sigma baryon $\Sigma_b{}^-$ (ddb) into a neutral bottom lambda baryon $\Lambda_b{}^0$ (udb) and a negative pion π^- (u`d):

$\Sigma_b{}^-$ (ddb) → $\Lambda_b{}^0$ (udb) + π^- (u`d)

In this decay, bottomness is conserved. Thus, this is a strong decay, i.e. through gluon.

In this decay, one **down quark** of the negative bottom sigma baryon **emits a gluon** and the down quark remains itself as a **down quark d**.
The **gluon then decays into an up quark u and an antiup quark u`**. This **antiup quark u`** combines with that **down quark d** of the negative bottom sigma baryon which had emitted the gluon and produces a **negative pion π^- (u`d)**.
The **up quark u** (produced by the decay of the gluon) combines with the the remaining down quark and the bottom quark of the negative bottom sigma baryon to produce a **neutral bottom lambda baryon $\Lambda_b{}^0$ (udb)**.

Charmed and Bottom Xi Baryons

1. The decay of the positive charmed Xi baryon Ξ_c^+ (usc) into a neutral Xi baryon Ξ^0 (uss), a positron e^+ and an electron-type neutrino v_e:

Ξ_c^+ (usc) $\rightarrow$ Ξ^0 (uss) + e^+ + v_e

In this decay, the **charm quark c** (charge: +2/3e) of the positive charmed Xi baryon **emits a W^+ boson** and **transforms into** a **strange quark s** (charge: -1/3e) and the positive charmed Xi baryon transforms into a **neutral Xi baryon Ξ^0 (uss)**.

The W^+ boson then decays into a **positron e^+** and an **electron-type neutrino v_e**.

2. The decay of the positive charmed Xi baryon Ξ_c^+ (usc) into a positive sigma baryon Σ^+ (uus), a negative kaon K^- (u`s) and a positive pion $\pi^+$ (ud`):

Ξ_c^+ (usc) $\rightarrow$ Σ^+ (uus) + K^- (u`s) + $\pi^+$ (ud`)

In this decay, the **charm quark c** (charge: +2/3e) of the positive charmed Xi baryon **emits a W^+ boson** and **transforms into** a **strange quark s** (charge: -1/3e).

The W^+ boson then decays into an **up quark u** (charge: +2/3e) and an **antidown quark d`** (charge: +1/3e) which combine to produce a **positive pion $\pi^+$ (ud`)**.

The **up quark** of the positive charmed Xi baryon **emits a gluon** and remains itself as an **up quark u**.

The **gluon then decays into an up quark u and an antiup quark u`**.

This **antiup quark u`** combines with the **strange quark s** of the positive charmed Xi baryon and produces a **negative kaon $K^-$ (u`s)**.

The **up quark u** (produced by the decay of the gluon) combines with that **up quark u** of the positive charmed Xi baryon which had emitted the gluon and the **strange quark s** (produced by the transformation of the charm quark) to produce a **positive sigma baryon Σ⁺ (uus).**

note: The antiup quark u` may also combine with the strange quark s (produced by the transformation of the charm quark) to produce a negative kaon K⁻ (u`s). In this situation, the **up quark u** (produced by the decay of the gluon) combines with that **up quark u** of the positive charmed Xi baryon which had emitted the gluon and the **strange quark s** of the positive charmed Xi baryon to produce a positive sigma baryon Σ⁺ (uus).

3. The decay of the neutral charmed Xi baryon Ξ_c^0 (dsc) into a lambda baryon Λ^0 (uds), an antineutral kaon $K^{0`}$ (d`s), a positive pion π^+ (ud`) and a negative pion $\pi^-$ (u`d):

Ξ_c^0 (dsc) → Λ^0 (uds) + $K^{0`}$ (d`s) + π^+ (ud`) + $\pi^-$ (u`d)

In this decay, the **charm quark c** (charge: +2/3e) of the neutral charmed Xi baryon **emits a W⁺ boson** and **transforms into a strange quark s** (charge: -1/3e).

The W⁺ boson then decays into an **up quark u** (charge: +2/3e) and an **antidown quark d`** (charge: +1/3e) which combine to produce a **positive pion $\pi^+$ (ud`).**

The **down quark** of the neutral charmed Xi baryon **emits a gluon** and remains itself as a **down quark d.**

The **gluon then decays into an up quark u and an antiup quark u`.**

This **antiup quark u`** combines with that **down quark d** of the neutral charmed Xi baryon which had emitted the gluon and produces a **negative pion $\pi^-$ (u`d).**

The **strange quark** of the neutral charmed Xi baryon also **emits a gluon** and remains itself as a **strange quark s.**

This second **gluon then decays into a down quark d and an antidown quark d`.** This **antidown quark d`** combines with that

strange quark s of the neutral charmed Xi baryon which had emitted the second gluon and produces an **antineutral kaon K⁰` (d`s)**.

The **up quark u** and the **down quark d** (emitted by the two gluons) combine with the **strange quark s** (produced by the decay of the W⁻ boson) and produces a **lambda baryon Λ⁰ (uds)**.

4. The decay of the neutral bottom Xi baryon Ξ$_b$⁰ (usb) into a proton p (uud), neutral D meson D⁰ (cu`) and a negative kaon K⁻ (u`s):

$$\Xi_b{}^0 \text{(usb)} \rightarrow \text{p (uud)} + \text{D}^0 \text{(cu`)} + \text{K}^- \text{(u`s)}$$

In this decay, the **bottom quark b** (charge: -1/3e) of the neutral bottom Xi baryon **emits a W⁻ boson** and **transforms into** a **charm quark c** (charge: +2/3e).

The W⁻ boson then decays into a **down quark d** (charge: -1/3e) and an **antiup quark u`** (charge: -2/3e). This **antiup quark u`** combines with the **charm quark c** (produced by the transformation of the bottom quark) and produces a **neutral D meson D⁰ (cu`)**.

The **strange quark** of the neutral bottom Xi baryon **emits a gluon** and remains itself as a **strange quark s**.

The **gluon then decays into an up quark u and an antiup quark u`**. This **antiup quark u`** combines with that **strange quark s** of the neutral bottom Xi baryon which had emitted the gluon and produces a **negative kaon K⁻ (u`s)**.

The **up quark u** (produced by the decay of the gluon) combines with that **up quark u** of the neutral bottom Xi baryon which had emitted the gluon and with the **down quark d** (produced by the decay of the W⁻ boson) and produces a **proton (uud)**.

5. The decay of the neutral bottom Xi baryon Ξ$_b$⁰ (usb) into a positive charmed lambda baryon Λ$_c$⁺ (udc) and a negative kan K⁻ (u`s):

Ξ_b^0 (usb) → Λ_c^+ (udc) + K^- (u`s)

In this decay, the **bottom quark b** (charge: -1/3e) of the neutral bottom Xi baryon **emits a W⁻ boson** and **transforms into** a **charm quark c** (charge: -1/3e +1e = +2/3e).

The W⁻ boson then decays into a **down quark d** (charge: -1/3e) and an **antiup quark u`** (charge: -2/3e). This **antiup quark u`** combines with the **strange quark s** of the neutral bottom Xi baryon and produces a **negative kaon K⁻ (u`s)**.

The **up quark u** of the neutral bottom Xi baryon, the **down quark d** (produced by the decay of the W⁻ boson) and the **charm quark c** (produced by the transformation of the bottom quark) combine to produce a **positive charmed lambda baryon Λ_c^+ (udc)**.

6. The decay of the negative bottom Xi baryon Ξ_b^- (dsb) into a negative Xi baryon Ξ^- (dss) and a J/ψ meson (cc`):

Ξ_b^- (dsb) → Ξ^- (dss) + J/ψ (cc`)

In this decay, strangeness and bottomness are not conserved. Thus, it is a weak decay, i.e. through W boson.

In this decay, the **bottom quark b** (charge: -1/3e) of the negative bottom Xi baryon **emits a W⁻ boson** and **transforms into** a **charm quark c** (charge: +2/3e).

The W⁻ boson then decays into a **strange quark s** (charge: -1/3e) and an **anticharm quark c`** (charge: -2/3e). This **anticharm quark c`** combines with the **charm quark c** (produced by the transformation of the bottom quark) and produces a **J/ψ meson (cc`)**.

The **strange quark s** (produced by the decay of the W⁻ boson) combines with the down quark and the strange quark of the negative bottom Xi baryon and produces a **negative Xi baryon Ξ^- (dss)**.

Charmed and Bottom Omega Baryons

1. The decay of the neutral charmed omega baryon Ω_c^0 (ssc) into an omega baryon Ω^- (sss) and a positive pion π^+ (ud`).

Ω_c^0 (ssc) $\rightarrow$ Ω^- (sss) + π^+ (ud`)

In this decay, strangeness and charmness are not conserved. Thus, this is a weak decay.

In this decay, the **charm quark c** (charge: +2/3e) of the neutral charmed omega baryon **emits a W⁺ boson** and **transforms into** a **strange quark s** (charge: -1/3e) and the neutral charmed omega baryon transforms into an **omega baryon Ω^- (sss).**

The W⁺ boson then decays into an **up quark u** (charge: +2/3e) and an **antidown quark d`** (charge: +1/3e) which combine to produce a **positive pion $\pi^+$ (ud`).**

2. The decay of the neutral charmed omega baryon Ω_c^0 (ssc) into an omega baryon Ω^- (sss), a positron e⁺ and an electron-type neutrino v_e.

Ω_c^0 (ssc) $\rightarrow$ Ω^- (sss) + e⁺ + v_e

In this decay, the **charm quark c** (charge: +2/3e) of the neutral charmed omega baryon **emits a W⁺ boson** and **transforms into** a **strange quark s** (charge: -1/3e) and the neutral charmed omega baryon transforms into an **omega baryon Ω^- (sss).**

The W⁺ boson then decays into a **positron e⁺** and an **electron-type neutrino v_e.**

3. The decay of the neutral charmed omega baryon Ω_c^0 (ssc) into an omega baryon Ω^- (sss), a positive pion π^+ (ud`) and a neutral pion $\pi^0$ (uu`).

Ω_c^0 (ssc) $\rightarrow$ Ω^- (sss) + π^+ (ud`) + $\pi^0$ (uu`)

In this decay, the **charm quark c** (charge: +2/3e) of the neutral charmed omega baryon **emits a W⁺ boson** and **transforms into** a **strange quark s** (charge: -1/3e).

The W⁺ boson then decays into an **up quark u** (charge: +2/3e) and an **antidown quark d`** (charge: +1/3e).

One **strange quark** of the neutral charmed omega baryon **emits a gluon** and remains itself as a **strange quark s.**

The **gluon then decays into an up quark u and an antiup quark u`.**

This **antiup quark u`** combines with the **up quark u** (produced by the decay of the W⁺ boson) and produces a **neutral pion π⁰ (uu`).**

The **up quark u** (produced by the decay of the gluon) combines with the **antidown quark d`** (produced by the decay of the W⁺ boson) and produces a **positive pion π⁺ (ud`).**

The two strange quarks of the neutral charmed omega baryon combine with the **strange quark s** (produced by the transformation of the charm quark) to produce an **omega baryon Ω⁻ (sss).**

4. The decay of the neutral charmed omega baryon Ω_c^0 (ssc) into an omega baryon Ω⁻ (sss), a negative pion π⁻ (u`d) and two positive pions π⁺ (ud`)

Ω_c^0 (ssc) → Ω⁻ (sss) + π⁻ (u`d) + π⁺ (ud`) + π⁺ (ud`)

In this decay, the **charm quark c** (charge: +2/3e) of the neutral charmed omega baryon **emits a W⁺ boson** and **transforms into** a **strange quark s** (charge: -1/3e).

The W⁺ boson then decays into an **up quark u** (charge: +2/3e) and an **antidown quark d`** (charge: +1/3e) which combine to produce a **positive pion π⁺ (ud`).**

One **strange quark** of the neutral charmed omega baryon **emits a gluon** and remains itself as a **strange quark s.**

The **gluon then decays into an up quark u and an antiup quark u`.**

The other **strange quark** of the neutral charmed omega baryon also **emits a gluon** and remains itself as a **strange quark s.**

This second **gluon then decays into a down quark d and an antidown quark d`.**

The **up quark u** (produced by the decay of the first gluon) combines with the **antidown quark d`** (produced by the decay of the second gluon) to produce a **positive pion $\pi^+$ (ud`).**

The **antiup quark u`** (produced by the decay of the first gluon) combines with the **down quark d** (produced by the decay of the second gluon) to produce a **negative pion $\pi^-$ (u`d).**

The two strange quarks of the neutral charmed omega baryon combine with the **strange quark s** (produced by the transformation of the charm quark) to produce an **omega baryon Ω^- (sss).**

5. The decay of the neutral charmed omega baryon Ω_c^0 (ssc) into a neutral Xi baryon Ξ^0 (uss), a negative kaon K^- (u`s) and a positive pion $\pi^+$ (ud`):

Ω_c^0 (ssc) $\rightarrow \Xi^0$ (uss) + K^- (u`s) + $\pi^+$ (ud`)

In this decay, strangeness and charmness are not conserved. Thus, this decay involves the weak process.

In this decay, the **charm quark c** (charge: +2/3e) of the neutral charmed omega baryon **emits a W$^+$ boson** and **transforms into** a strange quark s (charge: -1/3e).

The W$^+$ boson then decays into an **up quark u** (charge: +2/3e) and an **antidown quark d`** (charge: +1/3e) which combine to produce a **positive pion $\pi^+$ (ud`).**

One **strange quark** of the neutral charmed omega baryon **emits a gluon** and remains itself as a **strange quark s.**

The **gluon then decays into an up quark u and an antiup quark u`.**

This **antiup quark u`** combines with that **strange quark s** of the neutral charmed omega baryon which had emitted the gluon and produces a **negative kaon K⁻ (u`s)**.

The **up quark u** (produced by the decay of the gluon) combines with the **strange quark s** (produced by the transformation of the charm quark) and the remaining **strange quark s** of the neutral charmed omega baryon to produce a **neutral Xi baryon Ξ⁰ (uss)**.

6. The decay of the neutral charmed omega baryon Ω_c^0 (ssc) into a negative Xi baryon Ξ⁻ (uss), a negative kaon K⁻ (u`s) and two positive pions π⁺ (ud`):

Ω_c^0 (ssc) → Ξ⁻ (dss) + K⁻ (u`s) + π⁺ (ud`) + π⁺ (ud`)

In this decay, the **charm quark c** (charge: +2/3e) of the neutral charmed omega baryon **emits a W⁺ boson** and **transforms into** a **strange quark s** (charge: -1/3e).

The W⁺ boson then decays into an **up quark u** (charge: +2/3e) and an **antidown quark d`** (charge: +1/3e) which combine to produce a **positive pion π⁺ (ud`)**.

One **strange quark** of the neutral charmed omega baryon **emits a gluon** and remains itself as a **strange quark s**.

The **gluon then decays into an up quark u and an antiup quark u`**. This **antiup quark u`** combines with that **strange quark** of the neutral charmed omega baryon **which had emitted the gluon** and produces a **negative kaon K⁻ (u`s)**.

The other **strange quark** of the neutral charmed omega baryon also **emits a gluon** and remains itself as a **strange quark s**.

This second **gluon then decays into a down quark d and an antidown quark d`**. This **antidown quark d`** combines with the **up quark u** (produced by the decay of the first gluon) and produces the second **positive pion π⁺ (ud`)**.

The **down quark d** (produced by the decay of the second gluon) combines with that **strange quark s** of the neutral charmed omega baryon which had emitted the second gluon and with the **strange quark s** (produced by the transformation of the charm quark) and produces a **negative Xi baryon** Ξ^- **(dss)**.

7. The decay of the neutral charmed omega baryon Ω_c^0 **(ssc) into a positive sigma baryon** Σ^+ **(uus), two negative kaons K⁻ (u`s) and a positive pion π⁺ (ud`):**

Ω_c^0 **(ssc)** → Σ^+ **(uus) + K⁻ (u`s) + K⁻ (u`s) + π⁺ (ud`)**

In this decay, the **charm quark c** (charge: +2/3e) of the neutral charmed omega baryon **emits a W⁺ boson** and **transforms into** a **strange quark s** (charge: -1/3e).

The W⁺ boson then decays into an **up quark u** (charge: +2/3e) and an **antidown quark d`** (charge: +1/3e) which combine to produce a **positive pion π⁺ (ud`)**.

One **strange quark** of the neutral charmed omega baryon **emits a gluon** and remains itself as a **strange quark s**.

The **gluon then decays into an up quark u and an antiup quark u`**. This **antiup quark u`** combines with that **strange quark s** of the neutral charmed omega baryon which had emitted the gluon and produces **a negative kaon K⁻ (u`s)**.

The other **strange quark** of the neutral charmed omega baryon also **emits a gluon** and remains itself as a **strange quark s**.

This second **gluon then decays into an up quark u and an antiup quark u`**. This **antiup quark u`** combines with that **strange quark s** of the neutral charmed omega baryon which had emitted the second gluon and produces the second **negative kaon K⁻ (u`s)**.

Two **up quarks u** (produced by the decay of the two gluons) combine with the **strange quark s** (produced by the transformation of the charm quark) and produces a **positive sigma baryon Σ⁺ (uus)**.

8. The decay of the negative bottom omega baryon Ω_b^- (ssb) into an omega baryon Ω⁻ (sss) and a J/ψ meson (cc`):

Ω_b^- (ssb) → Ω⁻ (sss) + J/ψ (cc`)

In this decay, strangeness and bottomness are not conserved.

Strangeness S = -2 and -3 before and after the decay respectively.

Thus, this is a weak decay.

In this decay, the **bottom quark b** (charge: -1/3e) of the negative bottom omega baryon **emits a W⁻ boson** (and therefore emits charge: -1e) and **transforms into** a **charm quark c** (charge: -1/3e +1e = +2/3e). The W⁻ boson then decays into a **strange quark s** (charge: -1/3e) and an **anticharm quark c`** (charge: -2/3e). This **anticharm quark c`** combines with the **charm quark c** (produced by the transformation of the bottom quark) and produces a **J/ψ meson (cc`)**

The two strange quarks of the negative bottom omega baryon combine with the **strange quark s** (produced by the decay of the W⁻ boson) and produces a **negative omega baryon Ω⁻ (sss).**

Omega Baryons

Baryons with no up or down quarks are omega baryons. Negative omega baryon Ω^- (sss) consists of 3 strange quarks. If one or two or three quarks are heavy (c or b), one or two or three subscripts are used to identify it, e.g. Ω_{cc}^+ (scc), Ω_{ccc}^{++} (ccc).

The negative omega baryon or simply omega baryon Ω^- (sss) baryon decays only through the weak process and has therefore a relatively long mean lifetime: $\tau = 8.2 \times 10^{-11}$ s.

Neutral Charmed Omega Baryon Ω_c^0 (ssc) consists of 2 strange quarks and 1 charm quark.

(As the strange quark has electric charge: -1/3e and the charm quark has electric charge: +2/3e, thus neutral charmed omega baryon consisting of 2 strange quarks and one charm quark has electric charge: -1/3e - 1/3e + 2/3e = 0)

Negative Bottom Omega Baryon Ω_b^- (ssb) consists of 2 strange quarks and 1 bottom quark.

Decay modes of Ω^- (sss): Λ^0 (uds) + K^- (u`s)  (BR: 67.8%); $\Xi^0$ (uss) + $\pi^-$ (u`d) (BR: 23.6%); Ξ^- (dss) + π^0 (uu`) (BR: 8.6 %)

Discovery of Negative Omega Baryon Ω^- (sss)

The first omega baryon was discovered in 1964 in the bubble chamber. It was the negative omega baryon Ω^-, whose constituents were three strange quarks.

In those **first omega events**, incoming negative kaon K^- (u`s) collided with a proton p (uud) in the bubble chamber and produced a negative omega baryon $\Omega^-$ (sss), a positive kaon $K^+$ (us`) and a neutral kaon K^0 (ds`):

K^- (u`s) + p (uud) $\rightarrow$ $\Omega^-$ (sss) + $K^+$ (us`) + K^0 (ds`)

This is a strong interaction as strangeness is conserved. Strangeness is S = -1 before the interaction and S = -3 + 1+ 1 = - 1 after the interaction. In this case, **one up quark of the proton and the antiup quark of the negative kaon annihilate to a gluon which** then materializes into a **strange quark s** and an **antistrange quark s`**. Another up quark of the proton **emits a gluon** and remains itself as an up quark. The gluon subsequently decays into **a strange quark s and an anti-strange quark s`**. Thus, the two gluons involved in this interaction produce two strange quarks and two antistrange quarks. **These two strange quarks combine with the strange quark of the negative kaon and produces a negative omega baryon Ω^- (sss)**. One antistrange quark s` combines with that up quark u of the proton which had emitted the gluon and produces a positive kaon $K^+$ (us`). Another antistrange quark s` combines with the down quark d of the proton and produces a neutral kaon $K^0$ (ds`).

The negative omega baryon Ω^- (sss) decayed rapidly <u>in a cascade of decays</u> to lower mass particles:

1. **Ω^- (sss) $\rightarrow$ Ξ^0 (uss) + π^- (u`d)   (s $\rightarrow$ u through W boson)**
On the right side, first write uss, i.e. <u>one s of left replaced by u.</u> Close the bracket. Then write 0 as charge of uss is 0. Then write Ξ.

2. **Ξ^0 (uss) $\rightarrow$ Λ^0 (uds) + π^0 (uu`)   [in the decay of the Neutral Bottom Xi-star Baryon: $\Xi^-$ (dss) $\rightarrow$ $\Lambda^0$ (uds) + $\pi^-$ (u`d)]**
On the right side, first write uds, i.e. <u>one s of left replaced by d.</u> Close the bracket. Then write 0 as charge of uds is 0. Then write Λ.

3. **π^0 (uu`) $\rightarrow$ γ + γ**

4. **Λ^0 (uds) $\rightarrow$ p (uud) + π^- (u`d)**

5. **π^- (u`d) $\rightarrow$ $\mu^-$ + $v_\mu$`**

6. **$\mu^- \rightarrow e^- + v_e$` $+ v_\mu$**

The negative omega baryon Ω^- (sss) **decayed** into a neutral Xi baryon Ξ^0 (uss) and a negative pion π^- (u`d) after travelling some distance 24.6 mm in its lifetime $\tau = 8.21 \times 10^{-11}$ s:

$$\Omega^- \text{(sss)} \rightarrow \Xi^0 \text{(uss)} + \pi^- \text{(u`d)}$$

(Already explained in the topic: Examples of The Decays – Example 11. Explained again below)

In this decay, strangeness is not conserved. Strangeness is S = -3 and -2 before and after the decay respectively, thus this decay is through the weak process i.e. through W boson.

In this decay, **one of the strange quarks** (charge: -1/3e) of the negative omega baryon **emits a W⁻ boson** and **transforms into** an **up quark u** (charge: +2/3e) and the negative omega baryon transforms into a **neutral Xi baryon Ξ^0 (uss)**.

The W⁻ boson (charge: -1e) then decays into an **antiup quark u`** (charge: -2/3e) and a **down quark d** (charge: - 1/3e) thereby producing a **negative pion $\pi^-$ (u`d)**.

note: The neutral Xi baryon was **discovered** in laboratory as a daughter product from the decay of the negative omega baryon in 1964.

The neutral Xi baryon Ξ^0 (uss) then decayed into a lambda baryon Λ^0 (uds) and a neutral pion π^0 (uu`) after travelling some distance 87 mm in its lifetime $\tau = 2.9 \times 10^{-10}$ s:

$$\Xi^0 \text{(uss)} \rightarrow \Lambda^0 \text{(uds)} + \pi^0 \text{(uu`)}$$

(Already explained in the topic: Examples of The Decays – Example 7. Explained again below)

In this decay too, strangeness is not conserved. Strangeness is S = -2 and -1 before and after the decay respectively, thus this decay is through the weak process.

In this decay, one **strange quark** (charge: -1/3e) of the neutral Xi baryon **emits a W⁻ boson** and **transforms into** an **up quark u** (charge: +2/3e).

The W⁻ boson (charge: -1e) then decays into an **antiup quark u`** (charge: -2/3e) and a **down quark d** (charge: - 1/3e). This **antiup quark u`** combines with the **up quark u** (produced by the transformation of the strange quark) and produces a **neutral pion π⁰ (uu`).**

The **down quark d** (produced by the decay of the W⁻ boson) combines with the up quark and the surviving strange quark of the neutral Xi baryon to produce a **lambda baryon Λ⁰ (uds)**.

The **neutral pion π⁰ (uu`)** produced by the decay of the neutral Xi baryon Ξ⁰ **(uss)** decayed into photon-pair. The photon-pair **appeared** straight at the point, where the neutral Xi baryon **decayed** into Λ⁰ (uds) and π⁰ (uu`), i.e. where π⁰ (uu`) **produced** because neutral pion will travel almost zero distance in a lifetime $\tau = 8.5 \times 10^{-17}$ s, even if its velocity is nearly equal to that of light.

Each photon then decayed into e⁺e⁻ pair in traversing the liquid hydrogen. (Electromagnetic decay)

The lambda baryon Λ⁰ (uds) produced by the decay of the negative Xi baryon Ξ⁻ **(dss)** decayed into a proton p (uud) and a negative pion π⁻ (u`d) after travelling some distance 78.9 mm in its lifetime $\tau = 2.632 \times 10^{-10}$ s:

Λ⁰ (uds) → p (uud) + π⁻ (u`d)

(Already explained in the topic: Examples of The Decays – Example 4. Explained again below)

In this decay, strangeness is not conserved. Strangeness is S = -1 and 0 before and after the decay respectively, thus this decay is through the weak process that is, through a W boson.

In this decay, the **strange quark** (charge: -1/3e) of the lambda baryon **emits a W⁻ boson** and **transforms into** an **up quark u** (charge: +2/3e) and the lambda baryon transforms into a **proton p (uud).**

The **W⁻ boson subsequently decays into an antiup quark u`** **(charge: -2/3e) and a down quark d (charge: -1/3e).** The antiup quark and the down quark combine to produce a **negative pion π⁻ (u`d).**

The negative pion π⁻ (u`d) decays into a muon μ⁻ and a muon-type antineutrino $v_\mu$` after travelling some distance 7.8 meter in its life time τ = 2.6 ×10⁻⁸ s:

$$\pi^- (u`d) \rightarrow \mu^- + v_\mu`$$

(Already explained in the topic: Examples of The Decays – Example 2. Explained again below)

In this decay, the antiup quark u` (charge: -2/3e) and the down quark d (charge: -1/3e) of the negative pion **annihilate** to a W⁻ boson (charge: -1e) which then decays into a **muon μ⁻** and a **muon-type antineutrino $v_\mu$`.**

The **muon μ⁻** decays into an electron e⁻, an electron-type antineutrino v_e` and a muon-type neutrino v_μ after travelling some distance 660 meter in its life time τ = 2.2 ×10⁻⁶ s (c τ = 3 × 10⁸ m/s × 2.2 ×10⁻⁶ s = 660 m):

$$\mu^- \rightarrow e^- + v_e` + v_\mu$$

(Already explained in the topic: Examples of The Decays – Example 3. Explained again below)

In this decay, the muon μ⁻ with one unit of negative electric charge emits that one unit of negative electric charge by emitting a W⁻ boson and therefore transforms into the corresponding electrically neutral lepton that is, **muon-type neutrino v_μ.**

The W⁻ boson subsequently decays into an **electron e⁻** and the corresponding antineutrino, i.e. **electron-type antineutrino v_e`.**

note: In this decay, the lepton number is conserved. The muon μ^- having lepton number L = +1 decays into **e⁻** (L= +1), **v_e`** (L= -1) and **v_μ** (L= +1).

K⁺ (us`) produced along with the negative omega baryon Ω⁻ (sss) may decay as

K⁺ (us`) → μ^+ + v_μ

**K⁺ (us`)** → π⁰ (uu`) + π⁺ (ud`)

**K⁺ (us`)** → π⁰ (uu`) + e⁺ + v_e

**K⁺ (us`)** → π⁰ (uu`) + μ^+ + v_μ

**K⁺ (us`)** → π⁺ (ud`) + π⁺ (ud`) + π⁻ (u`d)

**K⁺ (us`)** → π⁺ (ud`) + π⁰ (uu`) + π⁰ (uu`)

K⁰ (ds`) produced along with** the negative omega baryon Ω⁻ (sss) **decays into π⁺ (ud`) and π⁻ (u`d)

The omega baryon Ω⁻ (sss) may also decay as follows:

Ω⁻ (sss) → Λ⁰ (uds) + K⁻ (u`s)

Λ⁰ (uds) → p (uud) + π⁻ (u`d)

**K⁻ (u`s):** → π⁻ (u`d) + π⁻ (u`d) + π⁺ (ud`)

Positive Bottom Sigma-star Baryon (uub)

Sigma means one strange quark and two up/down quarks. Bottom Sigma means one strange quark replaced by one bottom quark. So Bottom Sigma has no strange quark but one bottom quark. Charge of bottom quark is -1/3e. So Positive Bottom Sigma must have 2 up quarks (charge of up quark: +2/3e) as the second and the third quarks to make total charge +1e. So Positive Bottom Sigma-star Baryon means uub.

The positive bottom sigma-star baryon Σ^{*+}_b (uub) <u>decayed rapidly</u> to lower mass particles:

1. Σ^{*+}_b (<u>uub</u>) → Λ^0_b (<u>udb</u>) + π^+ (ud`) (decay through gluon) Σ and Λ each contain 2 up/down quarks

On the right side, first write udb i.e. one u of the left side replaced by d. Close the bracket. Then write 0 as the charge of udb is 0. Then write b. Then Λ.

2. Λ^0_b (udb) → Λ^+_c (udc) + π^- (u`d) (b → c through W boson)

On the right side, first write udc i.e. b of the left side replaced by c. Close the bracket. Then write + as the charge of udc is +1. Then write c. Then Λ.

3. Λ^+_c (udc) → p (uud) + K$^-$ (u`s) + $\pi^+$ (ud`)

In each of these three, there is a pion on the right side.

The positive bottom sigma-star baryon Σ^{*+}_b (uub) was **discovered** through its <u>rapid decay</u> into a neutral bottom lambda baryon Λ^0_b (udb) and a positive pion π^+ (ud`). $\Lambda^0_b$ (udb) and $\pi^+$ (ud`) **appeared** straight at the point, where the positive bottom sigma-star baryon Σ^{*+}_b (uub) **produced** because the positive bottom sigma-star baryon Σ^{*+}_b (uub) will travel almost zero distance in a **lifetime τ = 5.7 × 10^{-23} s:**

Σ*$_b$$^+$(uub) → Λ$_b$^{0} (udb) + π$^+$ (ud`)

(mass of **Σ*$_b$$^+$(uub)** = 5832.1 MeV, mass of **Λ$_b$^{0} (udb)** = 5619.5 MeV, mass of π$^+$ (ud`) = 139.57 MeV)

This decay occurs in the same way as Σ$_b$$^+$ (uub) → Λ$_b$^{0} (udb) + π$^+$ (ud`)

(Already explained in the topic: Charmed and Bottom Sigma Baryons – Example 4. Explained again below)

In this decay, bottomness is conserved. Thus, this is a strong decay, i.e. through gluon.

In this decay, one **up quark** of the positive bottom sigma-star baryon **emits a gluon** and the up quark remains itself as an **up quark u**.

The **gluon then decays into a down quark d and an antidown quark d`**. This **antidown quark d`** combines with that **up quark u** of the positive bottom sigma-star baryon which had emitted the gluon and produces a **positive pion π$^+$(ud`).**

The **down quark d** (produced by the decay of the gluon) combines with the remaining up quark and the bottom quark of the positive bottom sigma-star baryon and produces a **neutral bottom lambda baryon Λ$_b$^{0} (udb).**

The neutral bottom lambda baryon Λ$_b$^{0} (udb) decayed into a positive charmed lambda baryon Λ$_c$$^+$ (udc) and a negative pion π$^-$ (u`d) after travelling some distance 435 μm (micrometers) in its lifetime τ = 1.45 × 10^{-12} s:

Λ$_b$^{0} (udb) → Λ$_c$$^+$ (udc) + π$^-$ (u`d)

(Already explained in the topic: Charmed and Bottom Lambda Baryons – Example 3. Explained again below)

In this decay, bottomness and charmness are not conserved. Thus, this is a weak decay, i.e. through W boson.

In this decay, the **bottom quark b** (charge: -1/3e) of the neutral bottom lambda baryon **emits a W$^-$ boson** and **transforms into** a **charm quark**

c (charge: +2/3e) and the neutral bottom lambda baryon transforms into a **positive charmed lambda baryon Λ$_c$$^+$ (udc).**

The W$^-$ boson then decays into an **antiup quark u`** (charge: -2/3e) and a **down quark d** (charge: -1/3e) which combine to produce a **negative pion π$^-$ (u`d).**

The positive charmed lambda baryon Λ$_c$$^+$ (udc) decayed into a proton p (uud), a negative kaon K$^-$ (u`s) and a positive pion π$^+$ (ud`) after travelling some distance 60 μm (micrometers) in its lifetime τ = 2 × 10^{-13} s:

Λ$_c$$^+$ (udc) → p (uud) + K$^-$ (u`s) + π$^+$ (ud`)

(Already explained in the topic: Charmed and Bottom Lambda Baryons – Example 1. Explained again below)

In this decay, charmness and strangeness are not conserved. Thus, this is a weak decay, i.e. through W boson.

In this decay, the **charm quark c** (charge: +2/3e) of the positive charmed lambda baryon **emits a W$^+$ boson** and **transforms into** a **strange quark s** (charge: -1/3e).

The W$^+$ boson then decays into an **up quark u** (charge: +2/3e) and an **antidown quark d`** (charge: +1/3e) which combine to produce a **positive pion π$^+$ (ud`).**

The **up quark** of the positive charmed lambda baryon **emits a gluon** and the up quark remains itself as an **up quark u.**

The **gluon then decays into an up quark u and an antiup quark u`.**

This **antiup quark u`** combines with the **strange quark s** (produced by the transformation of the charm quark) and produces a **negative kaon K$^-$ (u`s).**

The **up quark u** (produced by the decay of the gluon) combines with the **down quark d** of the positive charmed lambda baryon and that **up**

quark u of the positive charmed lambda baryon which had emitted the gluon and produces a **proton p (uud).**

Neutral Bottom Xi-star Baryon (usb)

Xi means 2 strange quarks and one up/down quark. Bottom Xi means one strange quark replaced by one bottom quark. So Bottom Xi has one s and one b. Sum of the charge of one s and one b is -2/3e. So neutral Bottom Xi-star must have u (charge +2/3e) as the third quark to make total charge 0e. So Neutral Bottom Xi-star Baryon means usb.

The neutral bottom Xi-star baryon $\Xi^{*}_{b}{}^{0}$ (usb) decayed rapidly in a cascade of decays to lower mass particles:

1. $\Xi^{*}_{b}{}^{0}$ (usb) → $\Xi_{b}{}^{-}$ (dsb) + π⁺ (ud`) (decay through gluon)

On the right side, first write dsb, i.e. u of the left side replaced by d. Close the bracket. Then write – as charge of dsb is –1. Then write b. Then Ξ.

2. $\Xi_{b}{}^{-}$ (dsb) → Ξ^{-} (dss) + J/ψ (cc`) (b → c through W boson)

On the right side, first write dss, i.e. b of the left side replaced by s (NOT c). Close the bracket. Then write – as charge of dss is –1. Then write Ξ.

3. J/ψ (cc`) → μ⁺ + μ⁻

4. Ξ^{-} (dss) → Λ^{0} (uds) + π⁻ (u`d) (s → u through W boson)

5. Λ^{0} (uds) → p (uud) + π⁻ (u`d) (s → u through W boson)

This cascade of decays produced one proton, three charged pions (one positive and two negative) and two muons.

The existence of the neutral bottom Xi-star baryon $\Xi^{*}_{b}{}^{0}$ (usb) was established by detecting all of these particles and measuring the charge, momentum and point of creation (vertex) for each one.

The neutral bottom Xi-star baryon $\Xi^{*}_{b}{}^{0}$ (usb) was **discovered** in 2012, in the Large Hadron Collider LHC which accelerated proton-proton pairs to cms energy of 7000 GeV (i.e. each proton having energy of 3500 GeV).

The neutral bottom Xi-star baryon $\Xi^*_b{}^0$ (usb) was **discovered** through its rapid decay into a negative bottom Xi baryon Ξ_b^- (dsb) and a positive pion. Ξ_b^- (dsb) and π^+ (ud`) **appeared** straight at the point, where the neutral bottom Xi-star baryon $\Xi^*_b{}^0$ (usb) **produced** because $\Xi^*_b{}^0$ (usb) will travel almost zero distance in its **lifetime τ = 3.13 × 10⁻²² s:**

$\Xi^*_b{}^0$ (usb) → Ξ_b^- (dsb) + π^+ (ud`)

(mass of $\Xi^*_b{}^0$ (usb) = 5949.4 MeV, mass of Ξ_b^- (dsb) = 5794.9 MeV, mass of π^+ (ud`) = 139.57 MeV)

In this decay, strangeness and bottomness are conserved. Thus, this is a strong decay, i.e. through gluon.

When the three quarks - up, strange and bottom get together to produce a neutral bottom Xi-star baryon, they immediately separate because the bottom quark is very massive.

The **up quark** moves away from the strange and the bottom quark and **emits a gluon** and the up quark remains itself as an **up quark u.** The gluon then decays into a **down quark d** and an **antidown quark d`.** This **antidown quark d`** combines with that **up quark u** of the neutral bottom Xi-star baryon which had emitted the gluon and produces a positive pion π^+ **(ud`).** The **down quark d** (produced by the decay of the gluon) combines with the strange quark and the bottom quark of the neutral bottom Xi-star baryon and produces a **negative bottom Xi baryon Ξ_b^- (dsb)**

(The negative bottom Xi baryon Ξ_b^- (dsb) was the **first known particle made of quarks from** all three quark generations. The down, strange and bottom quarks are the first, second and third generation quarks respectively).

The negative bottom Xi baryon Ξ_b^- (dsb) decayed into a negative Xi baryon Ξ^- (dss) and a J/ψ meson (cc`) after travelling some distance 0.47 mm in its lifetime τ = 1.56 × 10⁻¹² s:

Ξ_b^- (dsb) → Ξ^- (dss) + J/ψ (cc`)

(Already explained in the topic: Charmed and Bottom Xi Baryons – Example 6. Explained again below)

In this decay, strangeness and bottomness are not conserved. Thus, it is a weak decay, i.e. through W boson.

In this decay, the **bottom quark b** (charge: -1/3e) of the negative bottom Xi baryon **emits a W⁻ boson** and **transforms into** a **charm quark c** (charge: +2/3e).

The W⁻ boson then decays into an **anticharm quark c`** (charge: -2/3e) and **strange quark s** (charge: -1/3e). This **anticharm quark c`** combines with the **charm quark c** (produced by the transformation of the bottom quark) and produces a **J/ψ meson (cc`).**

The **strange quark s** (produced by the decay of the W⁻ boson) combines with the down quark and the strange quark of the negative bottom Xi-star baryon and produces a **negative Xi baryon Ξ^- (dss).**

The J/ψ meson (cc`) produced by the decay of the negative bottom Xi baryon **Ξ_b^- (dsb)** decayed into a muon-antimuon pair μ⁺μ⁻. The muon-antimuon pair **appeared** straight at the point, where the negative bottom Xi baryon Ξ_b^- (dsb) **decayed** into Ξ^- (dss) and J/ψ meson, i.e. where J/ψ meson **produced** because J/ψ meson will travel almost zero distance in a lifetime τ = 7.1 × 10⁻²¹ s:

J/ψ (cc`) → μ⁺ + μ⁻

The **charm quark c** (charge: +1/3e) **and anti charm quark c`** (charge: -1/3e) **annihilate** to virtual photon which then decay into muon anti muon pair.

The negative Xi baryon Ξ^- (dss) produced by the decay of the negative bottom Xi baryon Ξ_b^- **(dsb)** decayed into a lambda baryon Λ^0 (uds) and a negative pion π^- (u`d) after travelling some distance 49.1 mm in its lifetime $\tau = 1.639 \times 10^{-10}$ s:

Ξ^- **(dss)** $\rightarrow$ Λ^0 **(uds)** + π^- **(u`d)**

(Already explained in the topic: Examples of The Decays – Example 6. Explained again below)

In this decay, strangeness is not conserved. Strangeness is S = -2 and -1 before and after the decay respectively, thus this decay is through the weak process.

In this decay, one **strange quark** (charge: -1/3e) of the negative Xi baryon **emits a W⁻ boson** and **transforms into** an **up quark u** (charge: +2/3e) and the negative Xi baryon transforms into a **lambda baryon Λ^0 (uds).**

The W⁻ boson (charge: -1e) then decays into an **antiup quark u`** (charge: -2/3e) and a **down quark d** (charge: -1/3e) which combine to produce a **negative pion $\pi^-$ (u`d).**

The lambda baryon Λ^0 (uds) produced by the decay of the negative Xi baryon Ξ^- **(dss)** decayed into a proton p (uud) and a negative pion π^- (u`d) after travelling some distance 78.9 mm in its lifetime $\tau = 2.632 \times 10^{-10}$ s:

Λ^0 **(uds)** $\rightarrow$ **p (uud)** + π^- **(u`d)**

(Already explained in the topic: Examples of The Decays – Example 4. Explained again below)

In this decay, strangeness is not conserved. Strangeness is S = -1 and 0 before and after the decay respectively, thus this decay is through the weak process that is, through a W boson.

In this decay, the **strange quark** (charge: -1/3e) of the lambda baryon **emits a W⁻ boson** and **transforms into** an **up quark u** (charge: +2/3e) and the lambda baryon transforms into a **proton p (uud).**

The **W⁻ boson subsequently decays into an antiup quark u`** **(charge: -2/3e) and a down quark d (charge: -1/3e).** The antiup quark and the down quark combine to produce a **negative pion π⁻ (u`d).**

Note:

The negative omega baryon Ω⁻ (sss) decayed rapidly <u>in a cascade of decays</u> to lower mass particles:

1. Ω^- (sss) $\rightarrow \Xi^0$ (uss) + π^- (u`d)
2. Ξ^0 (uss) $\rightarrow \Lambda^0$ (uds) + π^0 (uu`)
3. π^0 (uu`) $\rightarrow \gamma + \gamma$
4. Λ^0 (uds) $\rightarrow$ p (uud) + π^- (u`d)
5. π^- (u`d) $\rightarrow \mu^- + v_\mu$`
6. $\mu^- \rightarrow e^- + v_e$` + v_μ

The negative omega baryon Ω⁻ (sss) **decayed** into a neutral Xi baryon Ξ^0 (uss) and a negative pion π^- (u`d) after travelling some distance in its lifetime τ = 8.21×10^{-11} s.

The positive bottom sigma-star baryon Σ*ᵦ⁺ (uub) <u>decayed rapidly</u> to lower mass particles:

1. $\Sigma^*_b{}^+$ (<u>uub</u>) $\rightarrow \Lambda_b{}^0$ (<u>udb</u>) + π^+ (ud`)
2. $\Lambda_b{}^0$ (udb) $\rightarrow \Lambda_c{}^+$ (udc) + π^- (u`d)
3. $\Lambda_c{}^+$ (udc) $\rightarrow$ p (uud) + K^- (u`s) + $\pi^+$ (ud`)

The positive bottom sigma-star baryon Σ*ᵦ⁺ (uub) was **discovered** through its <u>rapid decay</u> into a neutral bottom lambda baryon $\Lambda_b{}^0$ (udb) and a positive pion π^+ (ud`). $\Lambda_b{}^0$ (udb) and $\pi^+$ (ud`) **appeared** straight at the point, where the positive bottom sigma-star baryon Σ*ᵦ⁺ (uub) **produced** because the positive bottom sigma-star baryon Σ*ᵦ⁺ (uub) will travel almost zero distance in a **lifetime τ = 5.7×10^{-23} s.**

The neutral bottom Xi-star baryon $\Xi^*_b{}^0$ (usb) decayed rapidly <u>in a cascade of decays</u> to lower mass particles:

1. $\Xi^*_b{}^0$ (usb) $\rightarrow \Xi_b^-$ (dsb) + π^+ (ud`)
2. Ξ_b^- (dsb) $\rightarrow \Xi^-$ (dss) + J/ψ (cc`)
3. J/ψ (cc`) $\rightarrow \mu^+ + \mu^-$
4. Ξ^- (dss) $\rightarrow \Lambda^0$ (uds) + π^- (u`d)
5. Λ^0 (uds) $\rightarrow$ p (uud) + π^- (u`d)

The neutral bottom Xi-star baryon $\Xi^*_b{}^0$ (usb) was **discovered** through its <u>rapid decay</u> into a negative bottom Xi baryon Ξ_b^- (dsb) and a positive pion. Ξ_b^- (dsb) and π^+ (ud`) **appeared** straight at the point, where the neutral bottom Xi-star baryon $\Xi^*_b{}^0$ (usb) **produced** because $\Xi^*_b{}^0$ (usb) will travel almost zero distance in its **lifetime $\tau = 3.13 \times 10^{-22}$ s.**

The Top Quark

The top quark t was discovered at the Fermi Lab proton-antiproton collider with 1800 GeV cms energy (i.e. the proton and the antiproton each having 900 GeV energy).

Only one in every 10 billions proton-antiproton (pp`) collisions produced a top-antitop quark pair. (If a bunch of protons having 50 billion protons is collided with a similar bunch of antiprotons, then 5 top-antitiop quark pairs will be produced).

During the electron-positron collisions at the cms energy of 1020 MeV, 3.1 GeV, 9.46 GeV, **a virtual photon is produced** in each of these cases which subsequently decays into a φ meson (strange-antistrange quark pair), a J/ψ meson (charm-anticharm quark pair), an ϒ meson (bottom-antibottom quark pair) respectively. Similarly, during the proton-antiproton collisions at the cms energy of 1800 GeV (the proton and the antiproton each having energy of 900 GeV) **a highly energetic gluon is produced which** subsequently decays into a top-antitop quark pair. (The top quark has rest mass 173.3 GeV and the top-antitop quark pair has mass of about 350 GeV).

Although colliding pp` pair has the cms energy of 1800 GeV, the entire cms energy is not converted into a gluon as pp` pair does not annihilate to a gluon.

(In case of e⁺e⁻ collision, the entire cms energy of e⁺e⁻ pair converts into a virtual photon to produce a vector meson).

In the proton-antiproton collision, the top-antitop pair may be produced through two ways:

1. One quark of the proton and one antiquark of the antiproton **annihilate** to a gluon (uu`→ g, dd` → g) of energy 350 GeV which subsequently decays into a top-antitop pair.

2. One quark of the proton and one antiquark of the antiproton each **emits** a gluon. These two gluons then **fuse** and decay into a top-antitop pair.

Mean lifetime of the top quark is about 5×10^{-25} sec which is 20 times shorter than the time scale for the strong interactions, hence the top quark does not have time to form hadrons (baryons or mesons) through the strong interaction. Thus, **there exists no baryon having top quark as its constituent and there exists no meson having top or antitop quark as its constituent.**

Also, the top quark (charge: +2/3e) can decay through the weak process only. It decays into a **real W⁺ boson** and a **bottom quark b** (charge: -1/3e). Similarly, the antitop quark t` (charge: -2/3e) of the top-antitop quark pair tt` decays into a **real W⁻ boson** and an **antibottom quark b`** (charge: +1/3e):

tt` → W⁺b + W⁻b`

The W⁺ boson then decays into a positron e⁺ and an electron-type neutrino v_e or decays into an antimuon μ⁺ and a muon-type neutrino v_μ:

W⁺ → e⁺ + v_e or μ⁺ + v_μ

The W⁻ boson then decays into an electron e⁻ and an electron-type antineutrino v_e` or decays into a muon μ⁻ and a muon-type antineutrino $v_\mu$`: W⁻ → e⁻ + v_e`or μ⁻ + $v_\mu$`

The bottom and the antibottom quarks fragment into hadrons.

Antibottom quarks produce positive B mesons B⁺ (ub`) too.

While the lifetime of both the top quark and the W bosons are unmeasurably short, mean lifetime of the positive B meson is т = 1.6×10^{-12} s, thus it travels some distance before decaying and may be identified as a secondary vertex separated from the primary vertex of the primary collisions of protons-antiprotons.

The top quark is unique in that, unlike all the other quarks, it is massive enough to decay into a real W⁺ boson.

note:

The top-antitop quark pair cannot be produced through the e^+e^- collision because to produce the top-antitop pair, the cms energy of colliding e^+e^- pair needs to be about or greater than 350 GeV (i.e. e^+ and e^- each having energy of 175 GeV), the mass of the top-antitop pair.

To accelerate an electron to the energy of 175 GeV, linac needs to be several kms long and bunches of the electrons and the positrons cannot be reused. **SLC (SLAC linear collider) can achieve the cms energy of 100 GeV for electron-positron pair.**

Also, cyclic accelerators cannot be designed to accelerate the electrons and the positrons to such an extent because of too much energy loss in the form of the synchrotron radiation.

Maximum cms energy for the electron-positron pair obtained even in the LEP circular collider was 209 GeV.

Thus, only pp or pp` colliders can be designed to achieve energy more than say, 100 GeV per particle.

V Particles

V particles referred to those massive particles (e.g. kaons, sigma baryons, lambda baryon) which decay into a pair of particles thereby producing 'V' pattern in a cloud or bubble chamber.

First 'V' events were observed in a cloud chamber in 1947 which included a neutral V event {due to what is now known as $K_S \rightarrow \pi^+ (ud`)$ + $\pi^- (u`d)$} and charged V events {due to what is now known as the decay $K^+ \rightarrow \mu^+ + v_\mu$}. As π^+ and π^- and similarly μ^+ and v_μ traversed in the cloud chamber in different directions, they formed 'V' track in the chamber.

In 1950, a neutral V-particle that had a proton (uud) and a negative pion (u`d) as its decay products was discovered. That decayed product was named lambda baryon or lambda hyperon Λ^0 (uds).
{Λ^0 (uds) $\rightarrow$ p (uud) + π^- (u`d)}
Rest mass of lambda baryon (1116 MeV) > rest mass of proton (938.272 MeV) + rest mass of negative pion (139 MeV).

Higgs Boson

The higgs boson H^0 was discovered in the Large Hadron Collider LHC which accelerated proton-proton pair to cms energy of 7000 GeV (i.e. each proton having energy of 3500 GeV).

Main mechanism for the production of a higgs boson in proton-proton collider is the gluon fusion that is, two of the gluons binding these protons collide and those highly energetic gluons form **virtual top or bottom quark loop which** then decays into a Higgs boson. (One gluon decays into a virtual top quark t and a virtual antitop quark t`. The virtual antitop quark t` absorbs the other gluon and remains itself as a virtual antitop quark t`. These virtual top quark and virtual antitop quark annihilate to a real Higgs boson).

The coupling of the Higgs boson to the quarks, i.e. the strength with which a quark can emit or absorb a higgs boson is proportional to the mass of the quarks. Thus, the heavier quarks, i.e. top and bottom quarks have the larger probability than the lighter quarks to decay into a higgs boson.

As the mass of the higgs boson is 125.7 GeV, it cannot decay into the top quark whose mass 173.3 GeV > 125.7 GeV. Also, the higgs boson may decay into one real Z boson and one virtual Z boson as the mass of a Z boson is 91 GeV < 125.7 GeV but Z boson pair have mass 182 GeV > 125.7 GeV.

Similarly, the higgs boson may decay into one real W boson and one virtual W boson.

Solar Neutrinos

Sources of neutrinos on the Earth surface are the Sun, Cosmic rays, Supernovae, etc.

Solar Neutrinos

The Sun's equatorial regions rotate the fastest. As you move toward the poles, the rotation slows down.

Solar rotation period is 25.67 days at the equator and 33.40 days at 75 degrees of latitude.

The motion of sunspots across the Sun's disk, noted by Galileo in the 17th century, provided the first evidence of the Sun's rotation.

When viewed from above the Sun's north pole, the Sun rotates counterclockwise.

Following reactions occur in the Sun:

$p + p \rightarrow {}^2H$ (1p, 1n) $+ e^+ + v_e + 0.42$ MeV $- (1)$

${}^2H$ (1p, 1n) $+ p \rightarrow {}^3He$ (2p, 1n) $+ \gamma + 5.51$ MeV $- (2)$

${}^3He$ (2p, 1n) $+ {}^3He$ (2p, 1n) $\rightarrow {}^4He$ (2p, 2n) $+ p + p + \gamma + 12.98$ MeV $- (3)$

$2*\{(1) + (2)\} + (3)$, i.e. $4p \rightarrow {}^4He$ (2p, 2n) $+ 2e^+ + 2v_e + 24.84$ MeV $- (4)$

${}^3He$ (2p, 1n) $+ p \rightarrow {}^4He$ (2p, 2n) $+ e^+ + v_e - (5)$

${}^3He$ (2p, 1n) $+ {}^4He$ (2p, 2n) $\rightarrow {}^7Be$ (4p, 3n) $+ \gamma - (6)$

${}^7Be$ (4p, 3n) $+ e^- \rightarrow {}^7Li$ (3p, 4n) $+ v_e + 0.861$ MeV or 0.383 MeV $- (7)$

${}^7Li$ (3p, 4n) $+ p \rightarrow {}^4He$ (2p, 2n) $+ {}^4He$ (2p, 2n) $- (8)$

${}^7Be$ (4p, 3n) $+ p \rightarrow {}^8B$ (5p, 3n) $+ \gamma - (9)$

${}^8B$ (5p, 3n) $\rightarrow {}^8Be^*$ (4p, 4n) $+ e^+ + v_e - (10)$

${}^8Be^* \rightarrow {}^4He + {}^4He - (11)$

These reactions are explained below.

The source of the radiation or radiation energy of the Sun (and other relatively lighter stars) is the proton-proton chain. In pp chain, a pair of protons (hydrogen nuclei) combine to produce a deuteron ^{2}H (1p, 1n) having one proton and one neutron, a positron e^+, and an electron-type neutrino v_e. In this reaction 0.42 MeV energy is released in the form of electron type neutrino v_e.

$p + p \rightarrow {}^2H \text{ (1p, 1n)} + e^+ + v_e + 0.42 \text{ MeV}$ – (1)

Total flux of neutrinos from the sun is dominated by this reaction but neutrinos from this **reaction** carry relatively low energy (0.42 MeV), thus **most detectors cannot detect them**.

Since the neutrinos interact weakly, they emerge unscathed while passing from the center to the surface of the sun.

{In the above reaction, in fact, a proton **p** converts into a neutron **n**, a positron e^+, and an electron-type neutrino v_e ($p \rightarrow n + e^+ + v_e$) **which is the reverse of beta decay ($n \rightarrow p + e^- + v_e`$).**

In a beta decay, a neutron transforms into a proton, an electron, and an electron antineutrino.

The positron e^+ emitted in this reaction is almost immediately annihilated with an electron e^- and their rest energy as well as kinetic energy is converted into the two photons γ. $e^- + e^+ \rightarrow 2 \text{ γ}$}

The deuteron ^{2}H (1p, 1n) soon combines with the another proton **p** to produce a helium-3 nucleus ^{3}He (2p, 1n) having 2 protons and 1 neutron. In this reaction 5.51 MeV energy is released in the form of γ.

$^2H \text{ (1p, 1n)} + p \rightarrow {}^3He \text{ (2p, 1n)} + \text{γ} + 5.51 \text{ MeV}$ – (2)

Thus in the formation of helium-3 nucleus ^{3}He by combing 3 protons, 0.42 MeV + 5.51 MeV (reactions 1 and 2) = 5.93 MeV energy is released and one positron e^+ and one electron-type neutrino v_e is produced.

$p + p + p \rightarrow$ ^{3}He + e$^+$ + v_e + γ + 5.93 MeV

The helium-3 nucleus ^{3}**He** (2p, 1n) combines with the another helium-3 nucleus ^{3}**He** (2p, 1n) to produce an alpha particle: the nucleus of helium ^{4}**He (2p, 2n),** and 2 protons. In this reaction, 12.98 MeV energy is released in the form of **γ**.

^{3}He (2p, 1n) + ^{3}He (2p, 1n) $\rightarrow$ ^{4}He (2p, 2n) + p + p + γ + 12.98 MeV – (3)

The overall reaction is **2*{(1) + (2)} + (3), i.e. 4p $\rightarrow$ ^{4}He (2p, 2n) + 2e$^+$ + 2v_e + 24.84 MeV - (4)**

Thus, **in the formation of an alpha particle by combining protons**, 2 * (0.42 MeV + 5.51 MeV) + 12.98 MeV = **24.84 MeV** energy is released and total **two** positrons e$^+$ and **two** electron-type neutrinos v_e are produced.

The annihilation of two positrons e$^+$ produces about 2.1 MeV, where 1.02 MeV energy is due to their rest energy and the remaining energy is due to their kinetic energy. Thus, total 24.84 + 2.1 ~ **26.9 MeV** energy is released.

In these reactions, **each electron-type neutrino v_e collects about 0.42 MeV,** thus two neutrinos collect about 0.84 MeV, and the rest goes to sunlight. **<u>Thus, for every 26 MeV of solar energy, two electron-type neutrinos v_e are produced.</u>**

Most of the energy of the supernova is released in the form of the neutrinos but only a small fraction of the energy of the sun is in the form of the neutrinos.

The **solar constant** is the energy obtained from the sun per minute per centimeter^2 area. The solar constant **at the Earth** is about 2 calories (or 52×10^{12} MeV) per minute per cm^2 and with each 26 MeV of solar energy, two electron-type neutrinos v_e reach the Earth. Thus, the **total electron-type neutrino flux at the Earth is about (52/13) × 10^{12} or 4 × 10^{12} electron-type neutrinos per minute per cm^2 or about 6 × 10^{10} electron-type neutrinos per second per cm^2** due to the reaction (4).

In addition to the reaction(1): $p + p \rightarrow {}^2H + e^+ + v_e + 0.42$ MeV, **neutrinos v_e may also be emitted by the reactions (5), (7) and (10) as follows :**

The helium-3 nucleus ^{3}**He** (2p, 1n) **combines** with another proton **p** to produce an alpha particle ^{4}**He (2p, 2n),** a positron **e$^+$**, and an electron-type neutrino **v_e.**

^{3}**He (2p, 1n) + p** $\rightarrow$ ^{4}**He (2p, 2n) + e$^+$ + v$_e$ – (5)**

{In the above reaction,, a proton **p** is converted into a neutron **n**, a positron **e$^+$** and an electron-type neutrino **v_e: (p** $\rightarrow$ n + e$^+$ + v$_e$)**

The helium-3 nucleus ^{3}**He** (2p, 1n) may also **combine** with an alpha particle (produced in the previous reactions, (3) or (5) to produce beryllium (4p, 3n), and **releases energy in the form of a photon:**

^{3}**He (2p, 1n) + ^{4}He (2p, 2n)** $\rightarrow$ ^{7}**Be (4p, 3n) + γ – (6)**

Beryllium (4p, 3n) captures an electron e$^-$ to produce lithium (3p, 4n) and an electron-type neutrino v_e. In this reaction, 0.861 MeV or 0.383 MeV energy is released in the form of an electron-type neutrino v_e:

^{7}**Be (4p, 3n) + e$^-$** $\rightarrow$ ^{7}**Li (3p, 4n) + v$_e$ + 0.861 MeV or 0.383 MeV – (7)**

{In the above reaction, a proton **p** (of beryllium) is converted into a neutron **n** (of lithium), and an electron-type neutrino v_e: $(p + e^- \rightarrow n + v_e)$

Each of the **90% neutrinos produced** by the conversion of beryllium to lithium **carry 0.861 MeV** (i.e lithium produced is in ground state) while each of the **remaining 10% carry 0.383 MeV** (i.e. lithium produced in the excited state).

Lithium (3p, 4n) **absorbs** a proton p to produce two alpha particles: (^{7}Li (3p, 4n) + p $\rightarrow$ ^{4}He (2p, 2n) + ^{4}He (2p, 2n) – (8)

Beryllium **may also absorb** a proton to produce boron (5p, 3n) and **releases energy in the form of a photon:**

^{7}Be (4p, 3n) + p $\rightarrow$ ^{8}B (5p, 3n) + γ – (9)

Boron (5p, 3n) <u>produces</u> an excited state of beryllium, i.e. **beryllium-8 (4p, 4n), a positron e$^+$ and an electron-type neutrino v_e:**

^{8}B (5p, 3n) $\rightarrow$ ^{8}Be* (4p, 4n) + e$^+$ + v_e – (10)

{In the above reaction, a proton p (of boron) is converted into a neutron n (of beryllium-8), a positron **e$^+$** and an electron-type neutrino v_e: $(p \rightarrow n + e^+ + v_e)$

Even though the electron-type neutrinos v_e produced by the conversion of boron-8 to beryllium-8 **are far less abundant, they carry high energy (up to 14.6 MeV) and most detectors are capable of detecting them.**

Beryllium-8 (4p, 4n) **decays** into two alpha particles: ^{8}Bc* $\rightarrow$ ^{4}He + ^{4}He – (11)

Atmospheric Neutrinos

The source of the atmospheric neutrinos is the cosmic rays. (See the topic: Cosmic Rays and Muons).

note: Solar neutrinos contain only electron type neutrinos v_e.

Atmospheric neutrinos contain electron-type neutrinos v_e, electron-type antineutrinos $v_e\grave{}$, muon-type neutrinos v_μ, muon-type antineutrinos $v_\mu\grave{}$.

Neutrino Oscillations

Neutrino Oscillation is a quantum mechanical phenomenon by which a neutrino of a particular flavour (electron-type, muon-type or tauon-type) **transforms into** a neutrino of different flavour as it propagates. For instance, a neutrino emitted as an electron-type neutrino v_e from the sun may **become** a muon-type neutrino v_μ during its propagation and hence may reach to the Earth surface as a muon–type neutrino v_μ.

The first evidence for the neutrino oscillations was obtained at **Kamiokande water detector** (precedessor to SuperK) in the early 1990s using atmospheric neutrinos, whose source is cosmic rays. Four reactions related to the Pion and Muon decays ($\pi^+ \rightarrow \mu^+ + v_\mu$, $\pi^- \rightarrow \mu^- + v_\mu$, $\mu^+ \rightarrow e^+ + v_e + v_\mu$ and $\mu^- \rightarrow e^- + v_e + v_\mu$) produce two v_μ , two v_μ one v_e and one v_e . This ratio holds in the low energy region: E_v < 1 GeV. Thus we expect that at sea level, we will get approximately **two muon-type neutrinos v_μ and two muon type antineutrinos v_μ for every electron-type neutrino v_e** and every electron type antineutrino v_e .

However, Kamiokande found roughly **equal number of** muon-type neutrinos v_μ and electron-type neutrinos v_e. This implied, some of the muon-type neutrinos v_μ **coming from the cosmic rays changed their flavour** to electron-type neutrinos v_e thereby decreasing the number of the muon-type neutrinos v_μ recorded and increasing the number of the electron-type neutrinos v_e recorded at the detector.

Kamiokande detector was able to sense the direction from which the neutrinos came, those from directly overhead which had travelled only 10 kms or so arrived in the expected ratio (2:1) but as the zenith angle increased (and therefore the distance travelled by muon-type neutrinos from upper atmosphere to the detector), the ratio decreased, i.e. the probability for a muon-type neutrino to convert into an electron-type neutrino was changing with the propagation of the muon-type neutrino. In 1998, the **SuperKamiokande water Cerenkov detector** confirmed the results of Kamiokande detector.

note: At higher energies, the ratio of muon μ^- to muon-type neutrino v_μ is larger, since a smaller fraction of muons undergo decay in flight in the atmosphere. The higher energy implies the larger velocity of muon and therefore the velocity of time decreases even more in the frame of the muon, i.e. lifetime of muon increases even more in the frame of the Earth, i.e. less μ^- will decay to $e^- + v_e` + v_\mu$ in flight.

The **SuperKamiokande water Cerenkov detector** consists of a cylinder capable of storing **50000 tons of water**. The surfaces of this detector are covered with 11000 photomultipliers which **record** the Cerenkov light produced by the relativistic particles (i.e. those moving with velocity nearly equal to that of light) traversing the dielectric medium. **As relativistic electrons traverse through water, they lose energy producing a cone of blue light (Cerenkov effect) and it is this light which is directly detected by the SuperKamiokande.**

Consider the interaction: $v_e + e^- \rightarrow v_e + e^-$.
When **an electron-type neutrino coming from the cosmic rays or the sun** enters the water filled inside Super Kamiokande, it collides with an electron in the water and undergoes **elastic neutrino-electron scattering ($v_e + e^- \rightarrow v_e + e^-$)**, and the outgoing electron is detected **by the Cerenkov radiation it emits in the water**.

In 2001, the Super Kamiokande experiment recorded 45% of the theoretically predicted number of total **solar neutrinos** but it cannot tell how much were v_e out of those 45%.

SuperKamiokande detector is sensitive to all types of neutrinos: v_e, v_μ, v_τ but the detection efficiency is 6.5 times greater for the electron-type neutrinos v_e than for the muon-type neutrinos v_μ and the tauon-type neutrinos v_τ.

Meanwhile, the Sudbury Neutrino Observatory (SNOLAB), the science facility in Canada, located 2100 meters underground, and consisting of **1000 tons of heavy water** also detected solar neutrinos.
The depth of the facility shields it from cosmic rays.

In heavy water too, **neutrinos can undergo elastic neutrino-electron scattering** after colliding with the electrons in heavy water which helps measure neutrinos flux.
In addition to this, there are two other interactions due to the neutrons present inside deuterons in heavy water: One is a charge current weak interaction and another is a neutral current weak interaction.

note:

A deuteron is the nucleus of a deuterium atom. Deuteron is composed of one proton and one neutron.
Deuterium means deuteron and an electron.
Deuterium (D_2O) is an isotope of hydrogen and has an atomic mass roughly twice that of hydrogen.
Deuterium nuclei i.e. deuterons fused with other protons and neutrons to create helium nuclei. Helium nucleus consists of 2 protons and 2 neutrons.

These newly formed helium nuclei then captured electrons from the surrounding plasma to create helium atoms. Helium atom consists of helium nucleus and two electrons.

Consider the interaction: $v_e + {}^2H$ (1p, 1n) $\rightarrow$ p + p + e⁻.

In charge current weak interaction which is actually **neutron absorption process**, when an **electron-type neutrino coming from the sun** enters the heavy water, it collides with a deuteron and emits W^+ boson and transforms into an electron and one of the down quark of the neutron (of a deuteron) absorbs that W^+ boson and transforms into u quark and in this way neutron transforms into proton : $v_e + {}^2H$ **(1p, 1n) $\rightarrow$ p + p + e⁻.**

Only electron-type neutrinos can participate in this interaction as explained below:

Maximum energy of an electron-type neutrino coming from the sun is less than 15 MeV whereas muon and tauon has rest mass 105 MeV and 1777 MeV respectively. This means if an electron-type neutrino coming from the sun **converts** into muon-type or tauon-type neutrino during propagation in the space, and then collides with deuteron in water, muon-type or tauon-type neutrino will have insufficient energy to convert into a muon or tauon by emitting W^+ boson.

In this interaction, the electron emitted carries off most of the neutrino's energy, on the order of 5 to15 MeV and is detected by the Cerenkov radiation it emits in heavy water.

This interaction enables one to measure the flux of the electron-type neutrinos v_e.

Using **neutrino absorption process** (which applies only to the electron-type neutrinos) SNO recorded 35% of the theoretically predicted number of total solar neutrinos and all those 35% were electron-type neutrinos v_e. So 65% solar neutrinos were missing. However, when **compared** with the SuperKamiokande data (45%), it appeared that 10% out of 45% neutrinos **detected** at SuperK must have been muon-type neutrinos v_μ or tauon-type neutrinos v_τ. However SuperK detector is 6.5 times more efficient for electron-type neutrinos v_e. this means, if the detector were as much efficient for v_μ or v_τ as it was for v_e, instead of 10%, 65% neutrinos **detected** at SuperK would have been muon-type neutrinos v_μ or tauon-type neutrinos v_τ. For example, if 45 neutrinos recorded in SuperK, then 35 were **v_e** and 10 were **v_μ or v_τ**, and as the SuperK detector is 6.5 times more efficient for electron-type neutrinos v_e, it means, if it had detected 10 v_μ or v_τ (along with 35 v_e), 65 v_μ or v_τ would be detected, if the detector were as much efficient for v_μ or v_τ as it was for v_e.

Thus, the **result of the data of these two different experiments** was that 35% neutrinos **received** from the sun were electron-type neutrinos and 65% neutrinos **received** from the sun were muon or tauon-type neutrinos and as only electron-type neutrinos are **emitted** by the sun, it means 65% electron-type neutrinos emitted by the sun **transformed** into muon or tauon-type neutrinos during propagation which was the evidence for the neutrino oscillation.

Neutrino oscillation solved the solar neutrino deficit problem that is, the observed number of the electron-type neutrinos coming from the sun were much less than predicted by theory but neutrino oscillation implied this was so because 65% electron-type neutrinos emitted by the sun **transformed** into muon or tauon-type neutrinos during propagation. Later on, SNO confirmed neutrino oscillations on its own.

In the neutral current weak interaction, a neutrino dissociates the deuteron into its constituents: neutron and proton. The neutrino continues on with slightly less energy, and all the three neutrino flavours are equally likely to participate in this interaction: **(v_e, v_μ, v_τ) + ^{2}H (1p, 1n) → n + p + (v_e, v_μ, v_τ).** In this interaction, a photon with roughly 6 MeV of energy is produced. This photon **collides** with an electron through Compton scattering and the **accelerated** electron can be detected through Cerenkov radiation emitted by it

This interaction enables one to measure the total flux of neutrinos i.e flux due to all three flavours of the neutrinos v_e, v_μ, v_τ.

Probability of, say, a muon type neutrino v_μ to become an electron-type neutrino v_e varies with the length L of propagation, i.e. P $(v_\mu \rightarrow v_e)$ = 1 - P $(v_\mu \rightarrow v_\mu)$, **where P $(v_\mu \rightarrow v_\mu)$ = 1 - sin^{2}2θ sin^2(1.27× Δm^{2}L/E).** Here Δm^2 = $m_\mu^2 - m_e^2$ (m_μ and m_e are the masses of muon and electron-type neutrinos).

As there does exist neutrino oscillation, there will be non-zero probability for v_μ to become v_e after some distance L travelled by muon-type neutrino, i.e. P $(v_\mu \rightarrow v_e)$ > 0 which implies **P $(v_\mu \rightarrow v_\mu)$ < 1 which** further implies **sin^2(1.27× Δm^{2}L/E) > 0, i.e.** 1.27× Δm^{2}L/E > 0 i.e Δm^2 = $m_\mu^2 - m_e^2$ > 0.

(As sin^{2}0 = 0, 1.27× Δm^{2}L/E can not be zero because this implies P $(v_\mu \rightarrow v_\mu)$ = 1)

Thus, even if v_e were massless, v_μ must have some mass. If mass of neutrinos were zero, i.e. Δm^2 = 0, then P $(v_\mu \rightarrow v_\mu)$ = 1 always, i.e. the muon-type neutrino would always remain as the muon-type neutrino while propagating through space and there were no oscillation. Thus, **neutrino oscillation is due to the non-zero mass of the neutrinos.**

Neutrinos are not massless but their masses are extremely smaller even with respect to the mass of the electron.

From all the available evidences, it can be said that the neutrino masses lie somewhere between 0.04 eV/c^2 and 0.4 eV/c^2.

Due to non-zero mass of neutrinos, there exists some cross generational mixing too.

The neutrinos are assumed to be massless and to exist in only one helicity state i.e LH helicity state. However, if the neutrino mass is small but finite, a tiny RH helicity component of neutrino will also exist.

KM3NeT (Cubic Kilometre Neutrino Telescope)

KM3NeT is a European research infrastructure with three sites: KM3NeT-It (ARCA), KM3NeT-Fr (ORCA), KM3NeT-Gr.

It is located at the bottom of the Mediterranean Sea.

It consists of two large neutrino detectors: ARCA and ORCA.

1. The ARCA (Astroparticle Research with Cosmics in the Abyss) neutrino detector is located in the Mediterranean Sea, about 80 kilometers (50 miles) off the coast of Portopalo di Capo Passero, Sicily, Italy, at a depth of roughly 3,500 meters. It is optimized for the detection of high-energy cosmic neutrinos in the TeV–PeV range.

One TeV (Tera-electron volt) is equal to 1000 GeV (Giga-electron volt).

One PeV (Peta-electron volt) is equal to 1000 TeV or 1 million GeV (Giga-electron volt).

The ARCA telescope is designed to have a detector volume of about 1 cubic kilometer.

The ARCA detector consists of 230 vertical string-like detection units, each with a height of about 700 meters.

These detection units are equipped with light sensor modules (optical modules) to detect the faint Cerenkov light generated by charged particles from neutrino interactions.

2. The ORCA (Oscillation Research with Cosmics in the Abyss) neutrino detector is located in the Mediterranean Sea, about 40 kilometers offshore Toulon, France, at a depth of 2450 meters. It is optimized for the detection of atmospheric neutrinos in the GeV range.

The ORCA is designed to study atmospheric neutrino oscillations and determine the neutrino mass hierarchy. The neutrino mass hierarchy refers to the ordering of the masses of the three neutrino flavors (electron, muon, and tauon).

The ORCA detector consists of 115 strings (Detection Units - DUs) each with a height of about 200 meters.

DUs are separated by 20 meters horizontally.

Each DU has 18 Digital Optical Modules (DOMs). DOMs are spaced 9 meters apart vertically. Each DOM contains 31 photomultipliers

3. KM3NeT-Gr is located in the Mediterranean Sea, approximately 20 km offshore Pylos, Peloponnese, Greece, at depths between 3500 and 5000 meters.
It is used for validation and qualification.

Seawater contains a small amount of heavy water, or deuterium oxide (D2O), with Vienna Standard Mean Ocean Water (VSMOW) containing about 156 deuterium atoms per million hydrogen atoms, meaning roughly 0.0156% of hydrogen atoms are deuterium.

Following interactions can occur inside the sea
1. $v_e + e^- \rightarrow v_e + e^-$
When **an electron-type neutrino coming from the distant astrophysical sources including cosmic rays** enters the sea, it collides with an electron in the water and undergoes **elastic neutrino-electron scattering ($v_e + e^- \rightarrow v_e + e^-$),** and the outgoing electron is detected **by the Cerenkov radiation it emits in the water.**
2. $v_e + {}^2H$ (1p, 1n) $\rightarrow p + p + e^-$

When an **electron-type neutrino coming from the distant astrophysical sources including cosmic rays** enters the sea, it collides with a deuteron in the water and emits W$^+$ boson and transforms into an electron and one of the down quarks of the neutron (of a deuteron) absorbs that W$^+$ boson and transforms into u quark and in this way neutron transforms into proton.

3. $v_\mu + {}^2H (1p, 1n) \rightarrow p + p + \mu^-$

Similarly, when a **muon-type neutrino coming from the distant astrophysical sources including cosmic rays** enters the sea, it collides with a deuteron in the water and emits W$^+$ boson and transform into a muon and one of the down quarks of the neutron (of a deuteron) absorbs that W$^+$ boson and transforms into u quark and in this way neutron transforms into proton.

These electrons and muons traversing the water are detected by the Cerenkov radiation they emit in the deep sea.

Arrays of thousands of optical sensors detect the Cerenkov light.

On February 13, 2023, ARCA detector of KM3NeT detected a cosmic neutrino of energy 220 PeV, the most energetic neutrino ever observed. This event is identified as KM3-230213A.

The recently discovered neutrino particles are 30 times more energetic than any other neutrino particles known so far.

White Dwarf and Type I Supernova

Giants and Supergiants

<u>Main sequence stars burn hydrogen into helium in their core.</u>

After the hydrogen-fusing period of a main-sequence star of low or medium mass ends, helium burning begins and star expands into a red giant.

After the hydrogen-fusing period of a main-sequence star of high mass ends, helium burning begins and star expands into a red supergiant.

<u>Red giants and red supergiants fuse helium into carbon and oxygen in their cores</u> by the triple-alpha process.

Triple alpha process

At sufficiently high temperatures (10^8 K) and densities, the **triple alpha process** can occur as follows:

$^4He + {}^4He \rightarrow {}^8Be + \gamma$

$^8Be + {}^4He \rightarrow {}^{12}C + \gamma$

That is, two apha particles (helium nuclei) fuse to form **unstable** beryllium. If another alpha particle can fuse with the beryllium nucleus before it decays, stable carbon is formed along with a gamma ray.

At even higher temperatures, other reactions can also occur by the capture of more alpha particles:

$^{12}C + {}^4He \rightarrow {}^{16}O + \gamma$ (at $6*10^8$ K)

$^{16}O + {}^4He \rightarrow {}^{20}Ne + \gamma$ (at 10^9 K)

Red supergiants fuse helium into carbon and oxygen at a faster rate, but during the periods of slow fusion (which means lesser outward radiation pressure), the star can contract in on itself and become a blue supergiant. They are blue because their temperature are spread over a smaller surface area making them hotter and blue in colour.

Red or blue supergiant may be massive enough to continue fusing heavier elements at its core until core consists of iron only. Then such a red or blue supergiant collapses, explodes as Type II supernova and becomes a neutron star.

In about five billion years, sun will become a red giant. This phase could last for about a billion years.

After the red giant phase, the Sun will shed its outer layers to form a planetary nebula and would itself become a white dwarf.

This white dwarf will slowly cool over time.

Red dwarfs have masses about 0.08 to 0.6 times that of the Sun. As the red dwarfs are less massive, they burn their hydrogen very slowly and efficiently. Thus **red dwarfs remain in the main sequence stage, i.e. burn hydrogen into helium for billions or even trillions of years.**

Thus they don't evolve into giant stars.

White Dwarf

Over billions of years, the entire helium in the core of a less massive red giants **converts into carbon and oxygen. Then the hot center core of the less massive <u>red giant</u> consists of carbon and oxygen** but is not sufficiently hot to fuse carbon and oxygen into heavier elements. Thus, the nuclear fusion in such a red giant ceases. This means the outward radiation pressure decreases over time and the inward gravitational force becomes more than outward radiation pressure. **<u>Such a star collapses and become a carbon oxygen white dwarf</u>. Thus, the core of the carbon oxygen white dwarf consists of carbon and oxygen.** If the mass of the star is about 10 solar masses, the core temperature will be sufficient to fuse carbon into neon and magnesium. In this case, an oxygen neon magnesium **(ONeMg) white dwarf or oxygen neon (ONe) white dwarf is formed.**

A white dwarf is very hot when it forms, but because it has no source of energy, it gradually cools as it radiates its energy away. This means that its radiation which initially has a high color temperature, will lessen and redden with time. Over a very long time, a white dwarf will cool and its material will begin to crystallize, starting with the core. The star's low temperature means it will no longer emit significant heat or light, and it will become a cold black dwarf. Because the length of time it takes for a white dwarf to reach this state is calculated to be longer than the current age of the known universe (approximately 13.8 billion years). it is thought that no black dwarfs yet exist. The oldest known white dwarfs still radiate at the temperatures of a few thousand <u>kelvins</u> which establishes an observational limit on the maximum possible age of the universe.

Type I Supernova

Type Ia supernova occurs due to the thermonuclear explosion of the core of the white dwarf (when the mass of the core exceeds Chandrasekhar limit i.e. 1.4 solar masses), usually leaves no remnant and lacks hydrogen lines but shows strong silicon lines **in its spectrum.**

Type Ia supernova is extremely luminous (L > 10^9 L$_{sun}$), even brighter than a Type II supernova.

If the mass of a white dwarf is less than the mass of the sun, then it is stable because the inward gravitational force is balanced by the outward pressure of electron degenerate gas. However, if the mass of the non-roatating white dwarf is greater than 10 to 15 solar masses, then the mass of the core of the white dwarf would be greater than 1.44 solar masses, the chandrashekar limit, and **if the mass of the core of the white dwarf is greater than 1.44 solar masses**, then the inward gravitational force becomes more than the outward pressure of electron degenerate gas. <u>**The core** (consisting of carbon and oxygen) **of such a white dwarf collapses.**</u> The collapse increases the temperature and density even further and the core undergoes thermonuclear fusion that is, a substantial fraction of the carbon and oxygen in the core of the white dwarf is converted into heavier elements within a period **of only a few seconds,** raising the core temperature to the billions of degrees. This **thermonuclear fusion causes the white dwarf to explode** violently and white dwarf becomes a Type Ia supernova. It releases a shock wave in which matter is typically ejected at speeds on the order of 5,000–20000 km/s, roughly 6% of the speed of light. **Due to the energy released in the explosion,** there is an extreme increase in luminosity. The typical absolute magnitude of Type Ia supernova is −19.3 (about 5 billion times brighter than the Sun).

Type Ia Supernova occurs in a binary systems in which one of the stars is a white dwarf. Material flows to the white dwarf from its larger companion.

T Coronae Borealis, also known as the Blaze Star or T CrB. **is a binary star system comprising a white dwarf and an ancient red giant about 3,000 light years away from Earth in the constellation** Corona Borealis that makes a distinctive "C" shape in the sky, primarily during the summer months.

The white dwarf which is the dead remnant of a star, is about the size of Earth but has the same mass as the sun. Meanwhile, the aging red giant is a dying star that's shedding material out into space. **The white dwarf's massive gravitational pull is hauling in the ejected material from the red giant. Once the white dwarf has accumulated enough material,** the core of the white dwarf collapses and undergoes thermonuclear fusion and the **white dwarf becomes a Type Ia supernova.**

The prior nova from this star system occurred in 1946. It's a cycle that's been repeating each 80 years since it was first discovered more than 800 years ago.

SN 1604, also known as Kepler's Supernova or Kepler's Star, was a Type Ia supernova that occurred in the Milky Way, in the constellation Ophiuchus. Appearing in 1604, it is the most recent supernova in the Milky Way galaxy to have been observed by the naked eye. During its highest luminosity, Kepler's Star was brighter than any other star in the night sky, with an apparent magnitude of −2.5. It was visible during the day for over three weeks.

Type II Supernova and Neutron Star

Type II supernova occurs due to the rapid collapse of the core of the massive star, leaves a neutron star or a black hole as the remnant and shows strong hydrogen lines in its spectrum. Progenitor stars for Type II supernova are usually red supergiants with masses ranging from 8 to 25 solar masses.

The source of the radiant energy of the stars is the nuclear binding energy released during the nuclear fusion of lighter elements into heavier elements. This fusion proceeds systematically through Periodic Table and the heavier elements are found successively in onion like layers with the heaviest nuclei (iron) in the hot center core.

note: In the case of the formation of the white dwarf – Over billions of years, the **entire helium** in the core of the less massive red giants **converts into carbon and oxygen.** In the case of the formation of the neutron star, following happens.

Over billions of years, all the **lighter elements** in the core of a **massive** star **convert into iron.**

Thus, the nuclear fusion in the massive star ceases. This means the outward radiation pressure decreases over time and the inward gravitational force becomes more than the outward radiation pressure. **The core** (consisting of iron) **of such a massive star collapses and undergoes neutronisation** that is, most of the iron nuclei in the core are fragmented into neutrons and protons and the Fermi energy of the electrons (the maximum energy of the fermions at zero temperature is called the Fermi energy) is enough (> 0.8 MeV) to initiate the conversion of proton into neutron and energy is released in the form of an electron-type neutrino: $e^- + p \rightarrow v_e + n$.

This process is called neutronisation. Due to this process, most of the protons in the core of the star are converted into neutrons. The collapsing core still contains iron nuclei, protons and electrons. However, the core of such a massive star now mainly consists of neutrons and electron-type neutrinos.

The neutron degeneracy pressure eventually prevents further collapse of the core and the collapsing **core** (consisting of neutrons and electron-type neutrinos) **of the massive star** 'bounces,' sending a powerful shockwave outwards.

Due to this, 10% out of total **neutrinos** produced through the process: $e^- + p \rightarrow v_e + n$ during neutronisation and carrying about 10% of the total gravitational energy release, **burst out** in a flash lasting few milliseconds, and the star explodes as a **Type II Supernova.**

However 90% of gravitational energy released during the neutronisation and carried mainly by electron-type neutrinos is temporarily locked in the core. **Even the most penetrating particles, the neutrinos, can only escape from within 100 meter or so of the surface.** Now, there is a thermal phase of the stellar core, in which neutrino-antineutrino pairs, electron-positron pairs and gamma rays will be in equilibrium. The remaining, i.e. 90% gravitational energy is emitted in the form of v_e, $v_e\grave{}$, v_μ, $v_\mu\grave{}$, v_τ, $v_\tau\grave{}$ **over several seconds** as the core cools down, by neutrino emission.

Thus, most of the supernova energy is released in the form of neutrinos (produced during the neutronisation) and **the supernova explosion leaves an extremely dense, small remnant made almost entirely of neutrons, called the neutron star.**

If the mass of the neutron star formed after supernova explosion is less than 2.2 to 2.9 solar masses then it is stable because the inward gravitational force is balanced by the outward pressure of neutron degenerate gas. The radius of such a neutron star is 10-12 kms. Neutron star does not produce radiation but its surface temperature can be 60,000°C. Instead of emitting light, neutron star releases energy in the form of neutrinos and cools down over time by neutrino emission.

A pulsar (short for 'pulsating star') is a rapidly spinning neutron star.

Neutron star emits high-energy beams at its North and South magnetic poles. If these beams are pointed at Earth, then as a Neutron star rotates, it seem to pulse. So, all Pulsars are Neutron stars, but not all Neutron stars are Pulsars.

Pulsar spins very fast, sometimes spinning hundreds of times a second (millisecond pulsars) and emits radio/X-rays pulses

Magnetic field strength of a pulsar is typically about 10^8 Tesla.

A magnetar is a rare type of neutron star with extremely strong magnetic field. Its surface magnetic field strength of a magnetar typically ranges from 10^9 to 10^{11} Tesla.

Magnetar spins slowly then a pulsar as the rotational energy is converted to magnetic energy and emits bursts of high-energy X-rays and gamma rays.

A stellar or stellar-mass black hole is a black hole that is formed when the core of the massive star collapses and explodes as Type II, Type Ib, Type Ic Supernova which leaves behind a neutron star.

If the progenitor star is much massive, then the neutron star formed after the supernova explosion may have mass exceeding the Tolman–Oppenheimer–Volkoff limit, approximately 2.2 to 2.9 solar masses and the inward gravitational force becomes more than the outward pressure of neutron degenerate gas and nuclear forces. **The core** (consisting of neutrons) **of such a neutron star collapses** and becomes a black hole. **Stellar black hole** has a mass 10 to 40 times the mass of the sun.

Stellar black holes and neutron stars are the remnants of the Type II, Type Ib or Type Ic supernova explosions.

White dwarfs are the stellar remnants i.e. the remnants of the stars after their death.

Neutrinos from Supernova 1987A

SN 1987A was a type II Supernova occurred outside Milky Way in LMC and was visible to the naked eye.

In all, about 10^{58} neutrinos were emitted from Type II supernova 1987A in the Large Magellanic Cloud and total about 10^{59} MeV energy was released **in a second or so**. At the Earth, about 163000 light years away, the flux of neutrinos (which had passed through the Earth on 23 February 1987) was over **10^{10} neutrinos through each square centimeter.** About 1% of the total energy released appeared in the form of electromagnetic radiation. The remaining energy was in the form of neutrinos.

An intense burst of neutrinos from Supernova 1987A had been detected in **multikilotonne** water Cerenkov detectors. This was the first experimental evidence that the most of the supernova energy is released in the form of neutrinos. 20 neutrinos from the Supernova 1987A had been detected in **Kamiokande and IMB** water Cerenkov detectors. 12 neutrinos each having energy greater than 6 MeV had been detected in Kamiokande, whereas 8 neutrinos each having energy greater than 20 MeV had been detected in IMB detector. The neutrino pulse arrived some seven hours before the optical signals became detectable.

Supernovae play a unique role in the production of the later part of the periodic table, since they are the only known sources of the extremely intense fluxes of neutrons which give rise to the rapid neutron capture chains that alone can build up the heavy elements.

The temperature in a Supernova can reach up to 1 billion degree celcius. This high temperature can lead to the production of heavy elements which may appear in the new nebula that results after the supernova explosion.

The iron in a person's blood, for example, was manufactured in Supernova explosions.

SN 1987A ejected 20000 Earth masses of radioactive iron.

Supernova may shine with the brightness of 10 billion Suns. The total energy released may be 10^44 joules, as much as the total energy released by the Sun during its 10 billion year lifetime.

Pair-Instability Supernova (PISN)

PISN occurs due to the thermonuclear explosion of the core of the extremely massive star.

As a very massive progenitor star (140 to 260 solar masses) evolves, its core reaches extremely high temperature (about 10^9 K) after carbon-fusing ends in the core. At this temperature, the core produces high-energy gamma rays. **These gamma rays provide the radiation pressure that supports the star against gravity. However, these high-energy gamma rays also begin to convert into electron-positron pairs and the radiation pressure decreases and** eventually the inward gravitational force becomes more than the outward radiation pressure. <u>**The core** (consisting of oxygen and silicon) **of such a massive star collapses. The collapse increases the temperature and density even further and the core undergoes thermonuclear fusion**</u> that is, a substantial fraction of the oxygen and silicon in the core of the star is converted into heavier elements within a period **of only a few seconds,** raising the core temperature to the billions of degrees. This **thermonuclear fusion causes the star to explode** violently and the star becomes a Pair-Instability Supernova.

Like Type Ia supernova, Pair-Instability Supernova leaves no remnant.

Kilonova

Kilonova occurs due to the merging of a neutron star with either a black hole or with another neutron star and create heavy elements like gold and platinum.

Formation of a **Double or Binary Neutron Star (DNS)** starts in a binary star system where the two massive stars orbit each other.

The more massive star collapses and explodes as a supernova and becomes a neutron star.

This newly formed neutron star pulls matter from its companion.

The companion star, having lost mass or evolved, also collapses and explodes as a supernova and becomes second neutron star.

Two stars orbiting a comman center of masses are known to emit gravitational waves.

Thus, **the two neutron stars orbiting each other lose energy by emitting gravitational waves and gradually spiraling inward and eventually collide and merge.**

The binary neutron star merger causes kilonova explosion which ejects neutron-rich material forming neutron-rich cloud where rapid neutron capture (r-process) form heavy elements like gold, platinum and uranium.

The radioactive decay of newly heavy element synthesized in rapidly cooling neutron-rich cloud emit electromagnetic radiation across the entire electromagnetic spectrum.

This bright flash of light lasting days to weeks is due to kilonova and is much brighter than a nova but dimmer than a supernova.

They are 10 to 100 times dimmer than supernovas. Nova is millions of times dimmer than supernova.

The first pulsar rotating in an orbit together with another neutron star was discovered in 1974 (Nobel Prize 1993).

GW170817 (Gravitational Waves), GRB170817A (Gamma-Ray Burst) and Kilonova AT 2017gfo

The source of GW170817, GRB170817A, Kilonova AT 2017gfo are the same cosmic event i.e. binary neutron stars merger located in the galaxy NGC 4993 about 140 million light-years away.

GW170817 is the gravitational wave signal from a binary neutron star merger. These signals were detected by LIGO/Virgo detectors. GRB170817A is the short gamma-ray burst from the same binary neutron star merger. GRB was detected by Fermi/INTEGRAL satellites. These messengers i.e. signals (gravity and light) had been detected on 17 August 2017.

The GW170817 (Gravitational signals) followed 1.7 seconds later by GW170817A (GRB), indicating that gravitational waves and light travel at nearly the same speed.

This observation also confirmed that the neutron stars merger produces GRB i.e. neutron stars merger is the progenitor of short GRBs.

The GW170817 and GW170817A (GRB) followed eleven hours later by longer-lasting radioactive afterglow visible across the electromagnetic spectrum from radio waves to X-rays and gamma-rays. This afterglow was due to kilonova **AT 2017gfo.** Kilonova detected by optical/infrared light (telescopes like Hubble).

The radioactive decay of newly heavy element synthesized in rapidly cooling neutron-rich cloud emitted electromagnetic radiation which peaked in optical and ultraviolet bands in less than a day, then faded quickly.

Spectral analysis of the kilonova AT 2017gfo confirmed neutron star mergers as a primary source for creating heavy elements like gold, platinum, and strontium through the rapid neutron capture process (r-process).

Kilonova i.e. the electromagnetic counterpart of GW170817 and GW170817A (GRB) allowed astronomers to pinpoint the merger's location in the galaxy NGC 4993, just 130 million light-years away.

Observations of the kilonova's afterglow continued for years, with X-ray emissions still detectable by the Chandra X-ray Observatory more than four years after the event.

Binary neutron stars merger is a multi-messenger event which emit gravitational waves (detectable by LIGO/Virgo), short gamma-ray bursts (detectable Fermi/Integral satellite) and optical/infrared light (detectable by telescopes like Hubble).

Binary neutron stars merger marked the beginning of multi-messenger astronomy.

Type Ib and Ic Supernova

Type Ib and Type Ic supernovae are core-collapse supernovae similar to Type II Supernovae. The key difference is their spectra.

Type II shows hydrogen lines, indicating the star still had its outer hydrogen layer before exploding.

Type Ib lacks hydrogen lines, indicating the star had lost its entire outer hydrogen before core collapse.

Type Ic lacks both hydrogen and helium lines, indicating the star lost its entire outer envelope (hydrogen and helium) before core collapse.

Progenitor stars for Type Ib supernovae have masses greater than 15 solar masses.

Progenitor stars for Type Ic supernovae have masses greater than 30 solar masses.

Hypernova

SNe occur due to the thermonuclear explosion of the core of the white dwarf (Type Ia) or the rapid collapse of the core of the massive star (Type II, Type Ib, Type Ic) or the thermonuclear explosion of a very massive star (PISN).

Hypernova is essentially an extremely energetic subset of Type Ic supernova, at least 10 times more luminous and energetic than Type II Supernova.

A small fraction of the Type Ic supernovae are the progenitors of long-duration gamma-ray bursts (GRBs). These Type Ic supernovae are often called hypernovae.

The core of the rapidly rotating massive star (> 30 solar masses) explodes as hypernova and collapses into black hole surrounded by an accretion disk. Near the event horizon of this new black hole, the accretion disk temperature may reach up to one hundred million kelvins. Matter at such high temperature emits high-energy thermal radiation, typically in the form of the **collimated jets of long gamma rays bursts (GRBs),** perpendicular to the accretion disk.

Thus, long-duration GRBs are typically associated with hypernovae (i.e. extremely energetic Type Ic supernovae).

Hydrogen-rich Type II supernova can occur without a detected GRB. However, hydrogen-rich Type II supernovae may also be associated with GRB-like phenomena, although it is rare. Example: **GRB 250314A** Hypernova is visible only if a jet is pointed towards Earth, appearing as a long-duration gamma-ray burst.

Duration of long GRB is > 2 seconds and up to 1000 seconds.

Short GRB occurs due to the mergers of ultra-dense objects like neutron stars or black holes. Duration of short GRB is < 2 seconds (milliseconds to 2s)

Duration of supernovs < days to weeks.

Supernova results in a neutron star or a black hole. Hypernova results in a black hole, often with an accretion disk and powerful jets.

GRB 250314A

GRB 250314A is a long-duration gamma-ray burst (GRB) detected on 14 March 2025 by the Space Variable Objects Monitor (SVOM) satellite.

Months later, JWST confirmed that this GRB event was due to the black hole formed after the Type II supernova explosion of a massive star. The explosion had hydrogen lines in its spectra which implied it could not be a hypernova which lacks hydrogen lines in its spectra.

GRB 250314A had occurred at a redshift of z ≈ 7.3, corresponding to 730 million years after the Big Bang i.e during the 'Epoch of Reionization (about 100 million to 1 billion years).'

GRB 090423 detected by the Swift Gamma-Ray Burst Mission had occurred at a redshift of z ≈ 8.26, corresponding to 630 million years after the Big Bang i.e during the 'Epoch of Reionization.'

Distant GRB like GRB 090423, GRB 250314A show that massive stars and galaxies existed in the very early universe.

Mean Lifetime

A typical mean lifetime τ for decay through a weak process is 10^{-10} seconds which is easily measurable while mean lifetime τ for a strong process is about 10^{-23} s which cannot be measured directly. However, an unstable particle does not have a unique mass but a distribution with width $\Gamma = \hbar/\tau$, so when τ is very short, its value can be deduced from the measured width Γ. Width Γ may be measured through the graph between cross-section versus cms energy of the colliding particles. In this way, we know the mean lifetimes of massive particles which decay quickly through the strong processes.

Planck constant h = 4.135×10^{-21} MeV-seconds
Reduced Planck constant $\hbar$ = h/2π = 6.582×10^{-22} MeV-s
For example, the **neutral rho meson ρ^0** has central mass 770 MeV and **width Γ = 150 MeV**. Thus, **mean lifetime of ρ^0** is $\tau = \hbar/\Gamma$ = 6.582 $\times$ 10^{-22} MeV-s/(150 MeV) = 0.044×10^{-22} s = **4.4×10^{-24} s.**
The **omega meson ω** has central mass 782 MeV and **width Γ = 8.5 MeV**. Thus, **mean life time of ω** is $\tau = \hbar/\Gamma$ = 6.582 $\times$ 10^{-22} MeV-s/(8.5 MeV) = 0.77×10^{-22} s = **7.7×10^{-23} s.**
The **J/ψ meson** has central mass 3.1 GeV and **width Γ = 92.9 KeV**. Thus, **mean lifetime of J/ψ** is $\tau = \hbar/\Gamma$ = 6.582 $\times$ 10^{-22} MeV-s/(92.9 $\times$ 10^{-3} MeV) = 0.071×10^{-19} s = **7.1×10^{-21} s.**

Masses and **Mean lifetimes** of some popular particles along with their **decay modes (and Branching ratios)** are as follows:

Nucleon
p (uud): m = 938.272046 MeV; $\tau > 10^{32}$ years; decay modes: **$e^+ + \pi^0$** (uu`); **$\mu^+ + \pi^0$ (uu`)**

n (udd): m = 939.565379 MeV; τ = 888.3 seconds; decay modes: **p (uud) + e⁻ + v_e`: β-decay** (Branching Ratio: ~ 100 %);  **p (uud) + e⁻ + v_e` + γ** (BR: 3.09×10^{-3} or 0.309 %)

Sigma Baryons

Σ⁺ (uus): m = 1189.37 MeV; τ = 8.018×10^{-11} s; decay modes: **p (uud) + π⁰ (uu`)**  (BR: 51.57%);  **n (udd) + π⁺ (ud`)** (BR: 48.31 %)

Σ⁰ (uds): m = 1192.642 MeV; τ = 7.4×10^{-20} s; decay mode: **Λ⁰ (uds) + γ** (BR: 100 %) (electromagnetic decay)

Σ⁻ (dds): m = 1197.449 MeV; τ = 1.479×10^{-10} s; decay modes: **n (udd) + π⁻ (u`d)**  (BR: 99.848 %);  **n (udd) + e⁻ + v_e`** (BR: 1.017×10^{-3} or 0.1017 %)

note: Σ⁺ (1189 MeV) cannot decay into Λ⁰ (1116 MeV) + π⁺ (139 MeV) as 1189 < 1116 + 139. Similarly, Σ⁰ (1192 MeV) cannot decay into Λ⁰ (1116 MeV) + π⁰ (134 MeV).

Lambda Baryon

Λ⁰ (uds): m = 1115.683 MeV; τ = 2.632×10^{-10} s; decay modes: **p (uud) + π⁻ (u`d)**  (BR: 63.9 %);  **n (udd) + π⁰ (uu`)** (BR: 35.8 %) (weak decay)

Xi Baryons

Ξ⁰ (uss): m = 1314.86 MeV; τ = 2.9×10^{-10} s; decay mode: **Λ⁰ (uds) + π⁰ (uu`)** (BR: 99.524 %)

Ξ⁻ (dss): m = 1321.71 MeV; τ = 1.639×10^{-10} s; decay mode: **Λ⁰ (uds) + π⁻ (u`d)** (BR: 99.887 %)

Delta Baryons

Δ⁺⁺ (uuu), Δ⁺ (uud), Δ⁰ (udd), Δ⁻ (ddd) or Δ(1232)
each has m = 1232 MeV; τ = 5.6×10^{-24} s;

Δ⁺⁺ (uuu): decay mode: **p (uud) + π⁺ (ud`)** (BR: 100 %)

Δ⁺ (uud): decay modes: **n (udd) + π⁺ (ud`)** or **p (uud) + π⁰ (uu`)** (BR: ~ 100 %); **p (uud) + γ** (BR: 0.55 %)

Δ⁰ (udd): decay modes: **n (udd) + π⁰ (uu`)** or **p (uud) + π⁻ (u`d)** (BR: ~ 100 %); **n (udd) + γ** (BR: 0.55 %)

Δ⁻ (ddd): decay mode: **n (udd) + π⁻ (u`d)** (BR: 100 %)

note: In all the decay modes (except those having BR < 1 %) of the delta baryons, strangeness is conserved which implies all these decays are through the gluons.

Sigma-star Baryons Σ*⁺ (uus), Σ*⁰ (uds), Σ*⁻ (dds)

Σ*⁺ (uus): m = 1382.80 MeV; τ = 1.8 × 10⁻²³ s; decay modes: **Λ⁰ (uds) + π⁺ (ud`)** (BR: 87 %);  **Σ⁺ (uus) + π⁰ (uu`)** or **Σ⁰ (uds) + π⁺ (ud`)** (BR: 11.7%); **Σ⁺ (uus) + γ** (BR: 0.7 %)

Σ*⁰ (uds): m = 1383.7 MeV; τ = 1.8 × 10⁻²³ s; decay modes: **Λ⁰ (uds) + π⁰ (uu`)** (BR: 87 %);  **Σ⁺ (uus) + π⁻ (u`d)** or **Σ⁰ (uds) + π⁰ (uu`)** (BR: 11.7%); **Λ⁰ (uds) + γ** (BR: 1.25 %)

Σ*⁻ (dds): m = 1387.2 MeV; τ = 1.6 × 10⁻²³ s; decay modes: **Λ⁰ (uds) + π⁻ (u`d)** (BR: 87 %);  **Σ⁰ (uds) + π⁻ (u`d)** or **Σ⁻ (dds) + π⁰ (dd`)** (BR: 11.7%); **Σ⁻ (dds) + γ** (BR: < 0.024 %)

note: In all the decay modes (except those having BR < 1 %) of the sigma-star baryons, strangeness is conserved which implies all these decays are through the strong processes that is, one of three quarks of a sigma-star baryon emits a gluon which then decays into quark-antiquark pair.

note: Σ*⁰ (1384 MeV) decays into Λ⁰ (1116 MeV) + π⁰ (134 MeV) as 1384 > 1116 + 134

The difference of energy (134 MeV) between initial states (1384 MeV) and final states (1116 + 134 =1250 MeV) is converted into the kinetic energy of the final states.

Xi-star Baryons

Ξ^{*0} (uss) or $\Xi(1530)^0$: m = 1531.8 MeV; τ = 7.2 × 10⁻²³ s; decay mode: Ξ^0 (uss) + π^0 (uu`) (BR: 100 %)

Ξ^{*-} (dss) or $\Xi(1530)^-$: m = 1535.0 MeV; τ = 6.6 × 10⁻²³ s; decay modes: Ξ^0 (uss) + π^- (u`d)  or  $\Xi^-$ (dss) + $\pi^0$ (dd`) (BR: 100 %)

note: All these decays of the Xi-star baryons are through the strong processes, as strangeness is conserved.

Omega Baryon Ω^- (sss):

m = 1672.45 MeV; τ = 8.21 × 10⁻¹¹ s; decay modes: Λ^0 (uds) + K⁻ (u`s)  (BR: 67.8%);  $\Xi^0$ (uss)  + $\pi^-$ (u`d) (BR: 23.6%); Ξ^- (dss) + π^0 (uu`) (BR: 8.6 %);

note: In all these three types of decays of the omega baryon, strangeness is not conserved which implies all these decays are through the weak processes.

Charged Kaons K⁺ (us`), K⁻ (u`s)

**K⁺ (us`):** m = 493.677 MeV;  $\tau$ = 1.238 × 10⁻⁸ s;  decay modes:  $\mu^+$ + $\nu_\mu$ (BR: 63.55 %);  $\pi^+$ (ud`) + π^0 (uu`)  (BR: 20.66 %);  $\pi^+$ (ud`) + π^+ (ud`) + $\pi^-$ (u`d) (BR: 5.59 %); π^0 (uu`) + e⁺ + $\nu_e$  (BR: 5.07 %);  $\pi^0$ (uu`) + μ^+ + ν_μ (BR: 3.35 %); π^+ (ud`) + $\pi^0$ (uu`) + π^0 (uu`) (BR: 1.76 %)

**K⁻ (u`s):** m = 493.677 MeV;  $\tau$ = 1.238 × 10⁻⁸ s;  decay modes:  $\mu^-$ + $\nu_\mu$` (BR: 63.55 %); π^- (u`d) + $\pi^0$ (uu`) (BR: 20.66 %); π^- (u`d) + $\pi^-$ (u`d) + π^+ (ud`)  (BR: 5.59 %);  $\pi^0$ (uu`) + e⁻ + ν_e`  (BR: 5.07 %);  $\pi^0$ (uu`) + μ^- + ν_μ`  (BR: 3.35 %);  $\pi^-$ (u`d) + π^0 (uu`) + $\pi^0$ (uu`) (BR: 1.76 %)

note: All these decays are weak decays, as strangeness is not conserved.

Neutral kaons K⁰ (ds`), K⁰` (or anti-K⁰) (d`s): m = 497.614 MeV

K^0 (ds`) and $K^{0`}$ (d`s) convert into Short-lived Kaon K_S and Long-lived Kaon K_L.

Neutral kaons K_S = $(K^0 + K^{0`})/2^{½}$  and  $K_L$  =  $(K^0 - K^{0`})/2^{½}$ (K^0 - ds`, $K^{0`}$ - d`s)

Short-lived Kaon K_S: m = 497.614 MeV; τ = 0.89 × 10^{-10} s; decay modes: π^+ (ud`) + $\pi^-$ (u`d) (BR: 69.20 %); π^0 (uu`) + $\pi^0$ (uu`) (BR: 30.69 %);

Long-lived Kaon K_L: m = 497.614 MeV; τ = 0.52 × 10^{-7} s; decay modes: π^+ (ud`) + $e^-$ + $v_e$` or π^- (u`d) + $e^+$ + $v_e$  (BR: 40.55 %);  $\pi^+$ (ud`) + μ^- + v_μ`  or  $\pi^-$ (u`d) + μ^+ + v_μ (BR: 27.04 %); π^0 (uu`) + $\pi^0$ (uu`) + π^0 (uu`)  (BR: 19.52 %);  $\pi^+$ (ud`) + π^- (u`d) + $\pi^0$ (uu`) (BR: 12.54 %)

Charged Pions π^+ (ud`), $\pi^-$ (u`d)

π^+ (ud`): m = 139.57018 MeV; τ = 2.6033 ×10^{-8} s; decay modes: μ^+ + v_μ (BR: 99.9877 %); e^+ + v_e (BR: 0.0123 %)

**π^- (u`d):**  m = 139.57018 MeV;  $\tau$ = 2.6033 ×$10^{-8}$ s;  decay modes:  $\mu^-$ + $v_\mu$` (BR: 99.9877 %); e^- + v_e` (BR: 0.0123 %)

Neutral Pion π^0 {(uu`- dd`)/$2^{1/2}$}:

m = 134.9766 MeV; τ = 8.52 × 10^{-17} s; decay modes: γ + γ (BR: 98.823 %), these two photons subsequently decay into e^+e^- pairs; e^+ + e^- + γ (BR: 1.174%)

note: The neutral pion π^0 decays about 10^9 rapidly as compared with its isospin partners (positive pions π^+ and negative π^-), this is because the neutral pion decays through the electromagnetic process and the charged pions decay through the weak processes.

Eta Meson η {(uu` + dd` - 2ss`)/$6^{1/2}$}:

m = 547.862 MeV; τ = 5 × 10⁻¹⁹ s; decay modes: **γ + γ** (BR: 39.41 %); **π⁰ (uu`) + π⁰ (uu`) + π⁰ (uu`)** (BR: 32.68 %); **π⁺ (ud`) + π⁻ (u`d) + π⁰ (uu`)** (BR: 22.92 %); **π⁺ (ud`) + π⁻ (u`d) + γ** (BR: 4.22 %); **e⁺ + e⁻ + γ** (BR: 0.69 %); **μ⁺ + μ⁻ + γ** (BR: 0.031 %)

Eta Prime Meson η′ {(uu` + dd` + ss`)/3¹ᐟ²} or η′(958):

m = 957.78 MeV; τ = 3.32 × 10⁻²¹ s; decay modes: **π⁺ (ud`) + π⁻ (u`d) + η** (BR: 42.9 %); **ρ⁰ + γ** (BR: 29.1%); **π⁰ (uu`) + π⁰ (uu`) + η** (BR: 22.2 %); **ω + γ** (BR: 2.75 %); **γ + γ** (BR: 2.20 %); **π⁰ (uu`) + π⁰ (uu`) + π⁰ (uu`)** (BR: 0.214 %);

Charged K-star Mesons K*⁺ (us`), K*⁻ (u`s)

K*⁺ (us`)** or **K*(892)⁺**: m = 891.66 MeV; τ = 1.3 × 10⁻²³ s; decay modes: **K⁺ (us`) + π⁰ (uu`)** or **K⁰ (ds`) + π⁺ (ud`) (BR: ~ 100%)
K*⁻ (u`s)** or **K*(892) ⁻**: m = 891.66 MeV; τ = 1.3 × 10⁻²³ s; decay modes: **K⁻ (u`s) + π⁰ (uu`)** or **K⁰` (d`s) + π⁻ (u`d) (BR: ~ 100%)

Neutral K-star Mesons K*⁰ (ds`), K*⁰` (or anti-K*⁰) (d`s)

K*⁰ (ds`):** m = 895.81 MeV; τ = 1.4 × 10⁻²³ s; decay modes: **K⁺ (us`) + π⁻ (u`d)** or **K⁰ (ds`) + π⁰ (dd`) (BR: ~ 100%)
K*⁰` (d`s): m = 895.81 MeV; τ = 1.4 × 10⁻²³ s; decay modes: **K⁻ (u`s) + π⁺ (ud`)** or **K⁰` (d`s) + π⁰ (dd`)** (BR: ~ 100%)
note: In all types of decays of charged and neutral K* mesons, strangeness is conserved which implies all these decays are through the strong processes that is, quark or anti-quark of a K-star meson (us`, u`s, ds` or d`s) emits a gluon which then decays into quark–antiquark pair: u and u` or d and d`.

Charged Rho Mesons ρ⁺ (ud`), ρ⁻ (u`d)

ρ^+ (ud`) or $\rho(770)^+$:  m = 775.26 MeV; $\tau$ = 4.4 × $10^{-24}$ s;  decay mode: $\pi^+$ (ud`) + π^0 (uu`) (BR: ~ 100%)

ρ^- (u`d) or $\rho(770)^-$:  m = 775.26 MeV; $\tau$ = 4.4 × $10^{-24}$ s;  decay mode: $\pi^-$ (u`d) + π^0 (uu`) (BR: ~ 100%)

Neutral Rho Meson ρ^0 {(uu`- dd`)/$2^{1/2}$} or $\rho(770)^0$:

m = 775.26 MeV τ = 4.4 × 10^{-24} s; decay mode: π^+ (ud`) + $\pi^-$ (u`d) (BR: ~ 100%)

note: All these decays of neutral and charged rho mesons are through the strong processes, as strangeness is conserved.

Omega Meson ω {(uu` + dd`)/$2^{1/2}$} or $\omega(782)$:

m = 782.65 MeV; τ = 7.7 ×10^{-23} s; decay modes: π^+ (ud`) + $\pi^-$ (u`d) + π^0 (uu`)  (BR: 89.2%);  $\pi^0$ (uu`) + γ (BR: 8.28%); π^+ (ud`) + $\pi^-$ (u`d) (BR: 1.53%)

Phi Meson φ (ss`) or $\varphi(1020)$:

m = 1019.46 MeV; τ = 1.54 × 10^{-22} s; decay modes: K^+ (us`) + $K^-$ (u`s) (BR: 48.9 %); K^0 (ds`) + $K^{0`}$ (d`s)  (BR: 34.2 %);  $\pi^+$ (ud`) + π^- (u`d) + $\pi^0$ (uu`) or ρ^+ (ud`) + $\pi^-$ (u`d) (BR: 15.32 %);

note: K^0 (ds`) and $K^{0`}$ (d`s) convert into Short-lived Kaon $K_S$ and Long-lived Kaon $K_L$, so $\varphi$ (ss`) $\rightarrow$ K_S + K_L (BR: 34.2 %)

J/Psi Meson J/ψ (cc`) or ψ (1S):

m = 3.096 GeV; τ = 7.1 × 10^{-21} s; decay modes: hadrons via 3 gluons (BR: 64.1 %); hadrons via virtual photon (BR: 13.5 %); hadrons via 2 gluons and 1 virtual photon (BR: 8.8 %); electron-positron pair e^+e^- (BR: 5.971 %); muon-antimuon pair $\mu^+\mu^-$ (BR: 5.961 %)

J/ψ (cc`) $\rightarrow$ hadrons (BR: 87.7)

ψ (2S):

m = 3.686 GeV; τ = 2.2 × 10^{-21} s; decay modes: **ψ (1S) + π$^+$(ud`) + π$^-$(u`d)** (BR: 34.45 %); **ψ (1S) + π^0(uu`) + π$^0$(uu`)** (BR: 18.13 %) **ψ (2S) → hadrons** (BR: 97.85)

ψ (3S) or ψ (3770):

m = 3.773 GeV; τ = 2.4 × 10^{-23} s; decay modes: **D^0 (cu`) + D$^{0`}$ (c`u)** (BR: 52 %);  **D$^+$(cd`) + D$^-$ (c`d)**  (BR: 41 %);  **J/ψ (cc`) + π$^+$(ud`) + π$^-$(u`d)** (BR: 1.93 %);

Upsilon Meson ϒ (bb`) or ϒ (1S):

m = 9.46 GeV; τ = 1.2 × 10^{-20} s; decay modes: hadrons via 3 gluons (BR: 81.7 %); hadrons via 2 gluons and 1 virtual photon (BR: 2.2 %); tauon-antitauon pair **τ$^+$τ$^-$** (BR: 2.60 %); muon-antimuon pair **μ$^+$μ$^-$** (BR: 2.48 %); electron-positron pair **e$^+$e$^-$** (BR: 2.38 %)

ϒ (2S):

m = 10.023 GeV; τ = 2.0 × 10^{-20} s; decay modes: **ϒ (1S) + π$^+$(ud`) + π$^-$(u`d)** (BR: 17.85 %); **ϒ (1S) + π^0(uu`) + π$^0$(uu`)** (BR: 8.6 %); **τ$^+$τ$^-$** (BR: 2.00 %); **μ$^+$μ$^-$** (BR: 1.93 %); **e$^+$e$^-$** (BR: 1.91 %) **ϒ (2S) → hadrons** (BR: 94 %)

ϒ (3S):

m = 10.355 GeV; τ = 3.2 × 10^{-20} s; decay modes: **ϒ (2S) + γ + γ** (BR: 5.00 %); **ϒ (2S) + π$^+$(ud`) + π$^-$(u`d)** (BR: 2.82 %); **ϒ (2S) + π^0(uu`) + π$^0$(uu`)** (BR: 1.85 %); **ϒ (1S) + π$^+$(ud`) + π$^-$(u`d)** (BR: 4.37)

ϒ (4S) or ϒ (10580):

m = 10.579 GeV; τ = 3.2 × 10⁻²³ s; decay modes: **B⁺ (ub`) + B⁻ (u`b)**

(BR: 51.4 %); **B⁰ (db`) + B⁰` (d`b)** (BR: 48.6 %)

D Mesons

D⁺ (cd`): m = 1869.61 MeV; τ = 1.04 × 10⁻¹² s; decay modes: **D⁰**

(cu`) + π⁺ (ud`) or **K⁰` (d`s) + π⁺ (ud`)**

D⁰ (cu`):** m = 1864.84 MeV; τ = 0.41 × 10⁻¹² s; decay mode: **K⁻ (u`s)

+ π⁺ (ud`)

Dₛ⁺ (cs`):** m = 1968.30 MeV; τ = 0.5 × 10⁻¹² s; decay mode: **φ (ss`) +

π⁺ (ud`)

B Mesons

B⁺ (ub`):** m = 5279.26 MeV; τ = 1.638 × 10⁻¹² s; decay mode: **D⁰`

(c`u) + π⁺ (ud`)

B⁰ (db`): m = 5279.58 MeV; τ = 1.519 × 10⁻¹² s; decay mode: **D⁻**

(c`d) + π⁺ (ud`)

Bₛ⁰ (sb`): m = 5366.77 MeV; τ = 1.512 × 10⁻¹² 4s; decay mode: **Dₛ⁻**

(c`s) + π⁺ (ud`)

B_c⁺ (cb`): m = 6275.60 MeV; τ = 0.452 × 10⁻¹² s; decay mode: **J/ψ**

(cc`) + π⁺ (ud`)

Top-antitop Quark Pair (tt`):

m = 350 GeV; τ = 5 × 10⁻²⁵ s; decay mode: **W⁺ b + W⁻b`**

Muon μ⁻:

m = 105.6583715 MeV; τ = 2.1969811 × 10⁻⁶ s; decay mode: **e⁻ + vₑ`**

+ v_μ (BR: 100 %)

Tauon τ⁻:

m = 1776.82 MeV; τ = 2.903 × 10⁻¹³ s; decay modes: **π⁻ (u`d) + v_τ** (BR: 64.82 %);  **e⁻ + v_e` + v_τ** (BR: 17.83 %); **μ⁻ + v_μ` + v_τ** (BR: 17.41 %);

W⁺, W⁻, Z⁰ Bosons

W⁺ Boson: m = 80.38 MeV; τ = 3.1 x 10⁻²⁵ s; decay modes: **e⁺ + v_e** (BR: 10.7 %); **μ⁺ + v_μ** (BR: 10.6 %); **τ⁺ + v_τ** (BR: 11.3 %); **π⁺ (ud`)**  or  **K⁺ (us`)** or **D⁺ (cd`)**  or  **D_s⁺ (cs`)** or **B⁺ (ub`)**  or  **B_c⁺ (cb`)** (BR: 67.4 %)

W⁻ Boson: m = 80.38 MeV; τ = 3.1 x 10⁻²⁵ s; decay modes: **e⁻ + v_μ`** (BR: 10.7 %);  **μ⁻ + v_μ`** (BR: 10.6 %); **τ⁻ + v_τ`**  (BR: 11.3 %);  **π⁻ (u`d)** or **K⁻ (u`s)**  or  **D⁻ (c`d)** or **D_s⁻ (c`s)**  or  **B⁻ (u`b)** or **B_c⁻ (c`b)** (BR: 67.4 %)

Z⁰ Boson: m = 91.18 MeV; τ = 2.6 x 10⁻²⁵ s; decay modes: **e⁺ + e⁻** (BR: 3.36 %); **μ⁺ + μ⁻** (BR: 3.36 %); **τ⁺ + τ⁻** (BR: 3.37 %); **hadrons** (BR: 69.9 %);

Higgs Boson H⁰:

m = 125.7 GeV; τ = 1.5 × 10⁻²² s; decay modes: bottom-antibottom pair **bb`** or **W⁺W⁻** or **2 Z⁰** or **2 gluons** or **2 photons** (2 γ) or tauon-antitauon pair **τ⁺τ⁻**

Positive Charmed Lambda Baryon Λ_c⁺ (udc):

m = 2286.46 MeV; τ = 2 × 10⁻¹³ s; decay mode: **Λ_c⁺ (udc) → p (uud) + K⁻ (u`s) + π⁺ (ud`)** (BR: 5%)

note: There are several decay modes of Λ_c⁺ (udc) and each of them as BR < 5% except p K⁻ π⁺.

Neutral Bottom Lambda Baryon Λ_b^0 (udb):

m = 5619.5 MeV; τ = 1.45 × 10^{-12} s; decay modes: Λ_b^0 (udb) → J/ψ (cc`) + $\Lambda^0$ (uds);  $\Lambda_b^0$ (udb) → $\Lambda_c^+$ (udc) + $\pi^-$ (u`d); Λ_b^0 (udb) → Λ_c^+ (udc) + D_s^- (c`s)

note: All these decays of Λ_b^0 (udb) are through the weak processes, as bottomness, charmness or strangeness is not conserved.

Double Positive Charmed Sigma Baryon Σ_c^{++} (uuc) or $\Sigma_c(2455)^{++}$:

m = 2453.98 MeV; τ = 2.91 × 10^{-22} s; decay mode: Σ_c^{++} (uuc) → Λ_c^+ (udc) + π^+ (ud`) (BR: ~ 100%)

Positive Charmed Sigma Baryon Σ_c^+ (udc) or $\Sigma_c(2455)^+$:

m = 2452.9 MeV; decay mode: Σ_c^+ (udc) → Λ_c^+ (udc) + π^0 (uu`) (BR: ~ 100%)

Neutral Charmed Sigma Baryon Σ_c^0 (ddc) or $\Sigma_c(2455)^0$:

m = 2453.74 MeV; τ = 3.04 × 10^{-22} s; decay mode: Σ_c^0 (ddc) → Λ_c^+ (udc) + π^- (u`d) (BR: ~ 100%)

Double Positive Charmed Sigma-star Baryon Σ_c^{*++} (uuc) or $\Sigma_c(2520)^{++}$:

m = 2517.9 MeV; τ = 4.41 × 10^{-23} s; decay mode: Σ_c^{*++} (uuc) → Λ_c^+ (udc) + π^+ (ud`) (BR: ~ 100%)

Positive Charmed Sigma-star Baryon Σ_c^{*+} (udc) or $\Sigma_c(2520)^+$:

m – 2517.5 MeV; decay mode: Σ_c^{*+} (udc) → Λ_c^+ (udc) + π^0 (uu`) (BR: ~ 100%)

Neutral Charmed Sigma-star Baryon Σ_c^{*0} (ddc) or $\Sigma_c(2520)^0$:

m = 2518.8 MeV; τ = 4.53 × 10^{-23} s;   decay mode: Σ_c^{*0} (ddc) → Λ_c^+ (udc) + π^- (u`d) (BR: ~ 100%)

Positive Bottom Sigma Baryon Σ_b^+ (uub)

m = 5811.3 MeV; τ = 6.8 × 10^{-23} s; decay mode: **Σ$_b^+$ (uub)** → **Λ$_b^0$ (udb) + π$^+$ (ud`)** dominant

Negative Bottom Sigma Baryon Σ$_b^-$ (ddb)

m = 5815.5 MeV; τ = 1.34 × 10^{-22} s; decay mode: **Σ$_b^-$ (ddb)** → **Λ$_b^0$ (udb) + π$^-$ (u`d)** dominant

Positive Bottom Sigma-Star Baryon Σ*$_b^+$ (uub):

m = 5832.1 MeV; τ = 5.7 × 10^{-23} s; decay mode: **Σ*$_b^+$ (uub)** → **Λ$_b^0$ (udb) + π$^+$ (ud`)** dominant

Negative Bottom Sigma-Star Baryon Σ*$_b^-$ (ddb):

m = 5835.1 MeV; τ = 8.7 × 10^{-23} s; decay mode: **Σ*$_b^-$ (ddb)** → **Λ$_b^0$ (udb) + π$^-$ (u`d)** dominant

Positive Charmed Xi Baryon Ξ$_c^+$ (usc):

m = 2467.8 MeV; τ = 4.42 × 10^{-13} s; decay modes: **Ξ$_c^+$ (usc)** → **Ξ^0 (uss) + e$^+$ + ν$_e$; Ξ$_c^+$ (usc)** → **Ξ^0 (uss) + π$^+$ (ud`); Ξ$_c^+$ (usc)** → **Σ$^+$ (uus) + K$^-$ (u`s) + π$^+$ (ud`);**

Neutral Charmed Xi Baryon Ξ$_c^0$ (dsc):

m = 2470.8 MeV; τ = 1.12 × 10^{-13} s; **Ξ$_c^0$ (dsc)** → **Λ^0 (uds) + K$^{0`}$ (d`s) + π$^+$ (ud`) + π$^-$ (u`d); Ξ$_c^0$ (dsc)** → **Λ^0 (uds) + K$^-$ (u`s) + π$^+$ (ud`) + π$^+$ (ud`) + π$^-$ (u`d)**

Positive Charmed Prime Xi Baryon Ξ′$_c^+$ (usc):

m = 2575.6 MeV; decay mode: **Ξ′$_c^+$ (usc)** → **Ξ$_c^+$ (usc) + γ**

Neutral Charmed Prime Xi Baryon Ξ′$_c^0$ (dsc):

m = 2577.9 MeV; decay mode: **Ξ′$_c^0$ (dsc)** → **Ξ$_c^0$ (dsc) + γ**

Positive Charmed Xi-star Baryon Ξ*$_c^+$ (usc) or Ξ(2645)$^+$:

m = 2645.9 MeV; decay mode: **Ξ*$_c^+$ (usc)** → **Ξ$_c^0$ (dsc) + π$^+$ (ud`)**

Neutral Charmed Xi-star Baryon Ξ*$_c^0$ (dsc) or Ξ(2645)^0:

m = 2645.9 MeV; decay mode: **Ξ*$_c^0$ (dsc)** → **Ξ$_c^+$ (usc) + π$^-$ (u`d)**

Neutral bottom Xi baryon Ξ_b^0 (usb):

m = 5793.1 MeV; τ = 1.49 × 10⁻¹² s; decay modes: Ξ_b^0 (usb) → p (uud) + D⁰ (cu`) + K⁻ (u`s); Ξ_b^0 (usb) → Λ_c^+ (udc) + K⁻ (u`s);

Negative bottom Xi baryon Ξ_b^- (dsb):

m = 5794.9 MeV; τ = 1.56 × 10⁻¹² s; decay mode: Ξ_b^- (dsb) → J/ψ (cc`) + Ξ⁻ (dss)

Neutral bottom Xi-star baryon $\Xi^*_b{}^0$ (usb) or Ξ_b(5945)⁰

m = 5949.4 MeV; τ = 3.13 × 10⁻²² s; decay mode: $\Xi^*_b{}^0$ (usb) → Ξ_b^- (dsb) + π⁺ (ud`)

Neutral Charmed Omega Baryon Ω_c^0 (ssc):

m = 2695.2 MeV; τ = 6.9 × 10⁻¹⁴ s; decay modes: Ω⁻ (sss) + π⁺ (ud`); Ω⁻ (sss) + e⁺ + $v_e$;  Ω⁻ (sss) + π⁺ (ud`) + π⁰ (uu`);  Ω⁻ (sss) + π⁻ (u`d) + π⁺ (ud`) + π⁺ (ud`); Ξ⁰ (uss) + K⁻ (u`s) + π⁺ (ud`); Ξ⁻ (dss) + K⁻ (u`s) + π⁺ (ud`) + π⁺ (ud`);  Σ⁺ (uud) + K⁻ (u`s) + K⁻ (u`s) + π⁺ (ud`)

Neutral Charmed Omega-star Baryon $\Omega^*_c{}^0$ (ssc) or Ω_c(2770)⁰:

m = 2765.9 MeV; decay mode: $\Omega^*_c{}^0$ (ssc) → Ω_c^0 (ssc) + γ (BR: ~ 100%)

Negative Bottom Omega Baryon Ω_b^- (ssb):

m = 6048.8 MeV; τ = 1.1 × 10⁻¹² s; decay mode: Ω_b^- (ssb) › J/ψ (cc`) + Ω⁻ (sss)

Electroweak Unification

Electroweak theory says, at energy ~ 100 GeV, the **coupling of W and Z bosons** to leptons and quarks should be the same as the **coupling of photons** to leptons and quarks, i.e. weak coupling and electromagnetic coupling are equal.

That is, **the electroweak unification energy is about 100 GeV**

At electroweak unification energy, $e^2 = 3g_w^2/8$ or $\alpha_e = 3\alpha_w/8$ or $\alpha_w = 2.66\,\alpha_e$.

That is, at energy about 100 GeV, the weak coupling constant is 2.66 times greater than the electromagnetic coupling constant. This implies, at energy about 100 GeV, the weak interaction is **intrinsically** 2.66 times stronger the electromagnetic interaction.

(note: As per supersymmetric grand unification, at electroweak unification, α_e ~ 1/63 and α_w ~ 1/26 and therefore α_w/α_e ~ 2.4)

However, at energy about 100 GeV, the effect of massive propagators W and Z bosons in the weak interaction/decay is such that decrease only about **2.66** times the strength of the weak interaction/decay, so that the weak interaction/decay (having weak coupling constant about 2.66 times greater than the electromagnetic coupling constant at electroweak unification energy) becomes almost **equal** in strength with respect to the electromagnetic interaction/decay. In this way, these two interactions get unified.

Even at the energy above electroweak unification, α_e continues to increase, α_w also continues to increase but much more slowly than α_e so that the ratio α_w/α_e approaches towards one and simultaneously the effect (which is already very small at electroweak unification) of the propagators in the strength of the weak interaction/decay continues to decrease and thus, both interactions remain unified even at energy much larger than 100 GeV.

Grand Unification Energy

The Supersymmetric (SUSY) version of grand unification implies **grand unification energy is about 3 × 10^{16} GeV.**

At the Grand Unification Energy, ratio α_w/α_e is indeed one.

That is, at 3 × 10^{16} GeV, the weak coupling constant and the electromagnetic coupling constants are equal and therefore the values of the weak and the electric charges are also the same.

Also, the effect of massive W and Z bosons in reducing the strength of the weak interactions no longer exists. Thus, both the interactions are of the same strength.

At 3 × 10^{16} GeV, not only are these two coupling constants the same, but also strong coupling contant is equal to these two coupling constants.

That is, at the Grand Unification Energy, the value of the strong, electromagnetic, weak coupling constants are equal and therefore the values of the strong, electric and weak charges are also the same.

Above the grand unification scale, there is just one universal coupling constant, and the strong, electromagnetic and weak interactions are identical in strength.

note:

At low energies, say, 1 MeV to 100 MeV:

e **= 0.3028**, α_e = $e^2/4\pi$ = $0.3028^2/4\pi$ = 0.0073 = **1/137**

g_w **= 0.65,** α_w = $g_w^2/4\pi$ = $0.65^2/4\pi$ = 0.034 = **1/29.5**

g_s **~ 2.96,** α_s = $g_s^2/4\pi$ ~ $2.96^2/4\pi$ ~ **0.7** ~ **1/1.43**

At low energies, say, 1 MeV to 100 MeV:

α_w/α_e = 137/29.5 = 4.64 i.e. α_w = **4.64** α_e

These values of coupling constants are true for phenomenon that occur at low energies. That is, for example, when the electron-positron pairs are collided at low cms energies, then the value of the coupling constants for various interactions will be as given above. However, when the electron-positron pairs are collided at cms energy say, 91GeV, the mass of the Z boson, then the value of the coupling constants for various interactions will be slightly different.

At E = 91 GeV:

$\alpha_e \sim$ 1/128 which is greater than 'α_e = 1/137 at low energy', i.e. α_e increases (very slowly) with the increase of energy-momentum transfer.

$\alpha_w \sim$ 1/29 which is greater than 'α_w = 1/29.5 at low energy', i.e. α_w increases (even more slowly then α_e) with the increase of energy-momentum transfer.

α_s = 0.117 = 1/8.5 which is smaller than '$\alpha_s \sim$ 0.7 at low energy', i.e α_s decreases with the increase of energy-momentum transfer.

At E = 91 GeV:

α_w / α_e = 128/29 = 4.41 i.e. α_w = **4.41 α_e**

At electroweak unification (~100 GeV):

$\alpha_e \sim$ 1/63

$\alpha_w \sim$ 1/26

$\alpha_s \sim$ 0.116

α_w / α_e = 63/26 = 2.42 i.e. **α_w = 2.42 α_e** (as per supersymmetric (SUSY) version of grand unification)

At grand unification: α_s (10^{16} GeV) = α_e (10^{16} GeV) = α_w (10^{16} GeV) ~ 1/24 (predicted)

Experimentally measured values of α_s at various energies are as follows:

At E = 1.58 GeV, α_s **(1.58 GeV)** = 0.375 = 1/2.67

At E = 1.78 GeV, α_s **(1.78 GeV)** = 0.323 = 1/3.09

α_s **(4.1 GeV)** = 0.239 = 1/4.18

α_s **(4.75 GeV)** = 0.217 = 1/4.60

α_s **(10.52 GeV)** = 0.20 = 1/5

α_s **(14 GeV)** = 0.170 = 1/5.88

α_s **(22 GeV)** = 0.151 = 1/6.62

α_s **(35 GeV)** = 0.145 = 1/6.89

α_s **(44 GeV)** = 0.139 = 1/7.19

α_s **(58 GeV)** = 0.132 = 1/7.57

α_s **(91.2 GeV)** = 0.117 = 1/8.54

α_s **(172 GeV)** = 0.104 = 1/9.61

Grand Unified Theory

In GUT, there are 45 fermions, i.e. both the leptons and the quarks which have the same universal coupling.

In GUT, there are 45 fermions, i.e. both the leptons and the quarks which have the same universal coupling.

The 45 fermions are split into three generations, each combining 15 states.

The first generation comprises the **up and down quarks,** each in three colours (red, blue, green) and two helicity states (LH and RH), the **electron** in two helicity states (LH and RH) and the **electron-type neutrino** in one helicity state (LH) only.

The particles in GUT are written down as LH states, the RH particles being replaced by LH antiparticles (since, by CP symmetry, the LH electron and the RH positron are equivalent).

Thus, for example, RH state of the up quark u having red colour is written as the LH state of the antiup quark u` having antired colour: $u`_R$ (charge: -2/3e).

The RH state of the down quark d having blue colour is written as the LH state of the antidown quark d` having antiblue colour: $d`_B$ (charge: +1/3e).

The RH state of electron is written as the LH state of positron: e^+ (charge: +1e).

The 15 particles of the first generation constitute following quintet and decuplet.

The quintet is: e^- (-1e), v_e (0e), $d`_R$ (+1/3e), $d`_B$ (+1/3e), $d`_G$ (+1/3e).

The decuplet is: e^+ (+1e), $u`_R$ (-2/3e), $u`_B$ (-2/3e), $u`_G$ (-2/3e), u_R (+2/3e), u_B (+2/3e), u_G (+2/3e), d_R (-1/3e), d_B (-1/3e), d_G (-1/3e).

note:

The subscripts R, B, G imply colour charge on a quark.

All these 15 particles in this representation are in LH helicity states.

The 15 particles of the second generation constitute following quintet and decuplet.

The quintet is: μ^- (-1e), v_μ (0e), s_R (+1/3e), s_B (+1/3e), s_G (+1/3e).

The decuplet is: μ^+ (+1e), c_R (-2/3e), c_B (-2/3e), c_G (-2/3e), c_R (+2/3e), c_B (+2/3e), c_G (+2/3e), s_R (-1/3e), s_B (-1/3e), s_G (-1/3e).

The 15 particles of the third generation constitute following quintet and decuplet.

The quintet is: τ^- (-1e), v_τ (0e), b_R (+1/3e), b_B (+1/3e), b_G (+1/3e).

The decuplet is: τ^+ (+1e), t_R (-2/3e), t_B (-2/3e), t_G (-2/3e), t_R (+2/3e), t_B (+2/3e), t_G (+2/3e), b_R (-1/3e), b_B (-1/3e), b_G (-1/3e).

In GUT, there are 12 known bosons, i.e. eight gluons, W^+, W^-, Z^0 bosons and a photon.

In addition to these bosons, there are massive bosons X and Y, with electric charges -4/3e and -1/3e respectively. These carry three colours (red, blue and green), and have antiparticles too. So X and Y bosons exist in 12 varieties.

Thus, there are 24 bosons in the GUT.

The 45 fermions interact through the mediation of these 24 bosons as explained below.

The X and Y bosons can couple <u>quarks to antileptons</u> (e.g. u → μ^+, d → e^+) and antiquarks to leptons (e.g. d` → e^-) that is, by emitting/absorbing an X, anti-X, Y or anti-Y boson, a quark can transform into an antilepton, and an antiquark can transform into a lepton. In this sense, these X, Y bosons are called the leptoquarks.

Transformation of a quark into an antilepton (q→ l`):

1. The up quark u (charge: +2/3e) may emit 1/3 unit of negative charge by **emitting a Y boson** (which carries 1/3 unit of negative charge: -1/3e) and transforms into an antimuon μ⁺ (charge: +2/3e + 1/3e = +1e). **(u → μ⁺).**

2. The down quark d (charge: -1/3e) may emit 4/3 unit of negative charge by **emitting a X boson** (which carries 4/3 unit of negative charge: -4/3e) and transforms into a positron e⁺ (charge: -1/3e + 4/3e = +1e). **(d → e⁺).**

Transformation of an antiquark into a lepton (q` → l):

3. The antidown quark d` (charge: +1/3e) may emit 4/3 unit of positive charge by **emitting an anti-X boson** (which carries 4/3 unit of positive charge: +4/3e) and transforms into an electron e⁻ (charge: +1/3e − 4/3e = -1e). **(d` → e⁻).**

The X and Y bosons can couple <u>quarks to antiquarks</u> **(u → u`, u → d`) and antiquarks to quarks (d` → u) too** that is, by emitting/absorbing an X, anti-X, Y or anti-Y boson, a quark can transform into an antiquark, and an antiquark can transform into a quark. In this sense, these X, Y bosons are called the diquarks.

Transformation of a quark into an antiquark (q → q`):

4. The up quark u (charge: +2/3e) may emit 4/3 unit of positive charge by **emitting an anti-X boson** (which carries 4/3 unit of positive charge: +4/3e) and transforms into an antiup quark u` (charge: +2/3 - 4/3 = -2/3e). **(u → u`).**

5. The up quark u (charge: +2/3e) may emit 1/3 unit of positive charge by **emitting an anti-Y boson** (which carries 1/3 unit of positive charge:

+1/3e) and transforms into an antidown quark d` (charge: +2/3 - 1/3 = +1/3e). (**u → d`**).

Transformation of an antiquark into a quark (q` → q):

6. The antidown quark d` (charge: +1/3e) may emit 1/3 unit of negative charge by **emitting a Y boson** (which carries 1/3 unit of negative charge: -1/3e) and transforms into an up quark u (charge: +1/3e + 1/3e = +2/3e). (**d` → u**).

Proton Decay

Just as the massive W and Z bosons make a weak decay very slow at low energies, similarly the decay rate of proton through the X, Y bosons is extremely rare, since the X and Y bosons are extremely massive. As per GUT, the mass of each of the X and Y boson is of the order of 10^{14} GeV. Grand unification implies that at energies about the masses of X any Y bosons (~ 10^{14} GeV), **a proton will transform into leptons via X, Y exchanges.** Even at normal energies, the virtual X, Y boson exchange can take place and although enormously suppressed on account of the massive propagators (X, Y bosons), will lead to the proton decay at some level.

note: If there are much more massive particles having mass on the order of M_{GUT}, these will inevitably occur in virtual processes at lower energy scales.

The leptoquark couplings allow for the nonconservation of lepton and baryon number and therefore allow the decay of the proton too. Thus, as per GUT, the proton is unstable and may decay for example, into a positron e^+ and a neutral pion π^0 (**uu`**):

p (uud) → e^+ + π^0 (uu`)

note: In proton decay, neither baryon number nor lepton number is conserved but the difference B − L is conserved, thus in grand unified theory, baryon number is not a conserved quantity.

The decay process may occur in many ways:

1. **p (uud)** → **e⁺ + π⁰ (uu`):** One up quark u (charge: +2/3e) of the proton **emits an anti-X boson** (charge: +4/3e) and transforms into an antiup quark u` (charge: -2/3e). This **antiup quark u`** combines with the other **up quark u** of the proton to produce a **neutral pion π⁰ (uu`).** The down quark d (charge: -1/3e) of the proton absorbs the anti-X boson (charge: +4/3e) and transforms into a **positron e⁺.** (p → e⁺ + π⁰).

2. **p (uud)** → **e⁺ + π⁰ (uu`):** The down quark d (charge: -1/3e) of the proton **emits an anti-Y boson** (charge: +1/3e) and transforms into an antiup quark u` (charge: -2/3e). This **antiup quark u`** combines with one **up quark u** of the proton to produce a **neutral pion π⁰ (uu`).** The other up quark u (charge: +2/3e) of the proton absorbs the anti-Y boson (charge: +1/3e) and transforms into a **positron e⁺.** (p → e⁺ + π⁰)

3. **p (uud)** → **e⁺ + π⁰ (dd`):** One up quark u (charge: +2/3e) of the proton **emits an anti-Y boson** (charge: +1/3e) and transforms into an antidown quark d` (charge: +1/3e). This **antidown quark d`** combines with the **down quark d** of the proton to produce a **neutral pion π⁰ (dd`).** The other up quark u (charge: +2/3e) of the proton absorbs the anti-Y boson (charge: +1/3e) and transforms into a **positron e⁺.**

4. **p (uud)** → **e⁺ + π⁰ (uu`):** The down quark d (charge: -1/3e) and one up quark u (charge: +2/3e) of the proton **annihilate to anti-Y boson** (charge: +1/3e) which subsequently decays into a **positron e⁺** and an antiup quark u` (charge: -2/3e). This **antiup quark u`** combines with the other **up quark u** of the proton to produce a **neutral pion π⁰ (uu`).**

In non SUSY unification, grand unification energy E_{GUT} is ~ 3 × 10^{14} GeV and universal coupling constant α_{GUT} = 1/43. Using these values, the lifetime of proton is calculated to be about 10^{31} years.

Although 10^{31} years is very long time, a kilotonne of material contains some 3 × 10^{32} protons. So 10^{31} years lifetime of proton implies a massive (kilotonne) detector would yield about one decay per 12 days per kilotonne of material. **To detect a proton decay, massive (kilotonne) detectors are placed deep underground** to reduce cosmic ray muon background. The largest of such detectors is the Superkamiokande detector which can detect the Cerenkov light that would be emitted, if relativistic positrons from the proton decay (p → e$^+$ + π^0) were to traverse the water.

As no proton decay has been detected in a kilotonne detector, it implied, by using various data from different detectors, **proton lifetime should be greater than 10^{32} years (т > 10^{32} years)** which is well above theoretically predicted value (т = 10^{30} to 10^{31} years).
Much larger expected proton lifetime implies much higher value of grand unification energy and the Supersymmetric (SUSY) version of grand unification implies **grand unification energy is about 3 × 10^{16} GeV with universal coupling constant α_{GUT} = 1/24 and the mass of each of the X and Y bosons is about 3 × 10^{16} GeV.**

The Super-Kamiokande collaboration has found that the proton lifetime for a specific decay pathway is greater than 5.9 x 10^{33} years.
The maximum upper limit on proton lifetime (if unstable) is calculated at 6×10^{39} years for SUSY-GUTs and 1.4×10^{36} years for non-SUSY GUTs. If protons were to decay, it would mean that all the matter in the universe would eventually disintegrate.

At grand unification energy E_{GUT} = 3 × 10^{16} GeV, strong, electromagnetic, weak forces are assumed to be unified that is, α_s = α = α_w or g_s = e = g_w that is, the **coupling** of gluon, photon, W, Z bosons to leptons and quarks are the same that is, the **strength** with which quarks (carrying strong, electric and weak charges) emit or absorb gluon, photon, W, Z bosons, the **strength** with which the charged leptons (carrying electric and weak charges) emit or absorb photon, W, Z bosons and the **strength** with which the neutrinos (carrying only weak charges) emit or absorb W, Z bosons, are the same.

Supersymmetry

According to the supersymmetry (SUSY), in addition to 45 fermions and 24 bosons of GUT, there are even more new fundamental fermions and bosons.

SUSY partner of a quark Q is called the squark Q˜, i.e. SUSY partners of up quark u, down quark d, strange quark s, charm quark c, bottom quark b, top quark t are up squark u˜, down squark d˜, strange squark s˜, charm squark c˜, bottom squark b˜, top squark t˜ respectively.

SUSY partner of a lepton l is called the slepton l˜, i.e. SUSY partners of electron e^-, muon μ^-, tauon τ^-, electron-type neutrino v_e, muon-type neutrino v_μ, tauon-type neutrino v_τ are selectron $e^-{}^˜$, smuon $\mu^-{}^˜$, stauon $\tau^-{}^˜$, electron-type sneutrino $v_e{}^˜$, muon-type sneutrino $v_\mu{}^˜$, tauon-type sneutrino $v_\tau{}^˜$ respectively.

The quarks and the leptons are spin 1/2 particles, whereas squarks and sleptons are spin 0 particles.

The SUSY partner of the photon γ (spin 1), gluon G (spin 1), W⁺, W⁻, Z⁰ bosons (spin 1) are photino γ˜ (spin 1/2), gluino G˜ (spin 1/2), winos W⁺˜, W⁻˜ (spin 1/2) and zino Z⁰⁻(spin 1/2) respectively.

Supersymmetry implies every fermion has a bosonic partner and every boson has a fermionic partner that is, **squarks and sleptons are bosons** whereas **photino, gluino, wini and zino are ferminos.**

Each SUSY particle is assumed to have a special quantum number R = + 1 and its anti-SUSY particle has R = -1 and in any interaction, this quantum number is conserved just as strangeness is conserved in strong interactions.

Thus, in quark-antiquark annihilation, a squark-antisquark pair can be produced $Q + Q` \rightarrow Q\tilde{} + Q\tilde{}`$, so that quantum number R is conserved. Each SUSY particle would decay in an R-conserving cascade process or directly to the lightest superparticle which will be stable. If this superparticle were the photino $\gamma\tilde{}$, the production of a squark would be manifest in the decay $Q\tilde{} \rightarrow Q + \gamma\tilde{}$.

Supersymmetry could not be exact, otherwise the superpartners would have the same mass as the original particles. For example, photino (a fermion) having spin 1/2 would then be a massless particle (like photon) and selectron (a boson) having spin 0 would then be a particle with a mass of 0.511 MeV (like electron) but we know, there is no existence of these particles.

Thus, this is a broken symmetry. The particles and their supersymmetric (SUSY) partners are assumed to have the same couplings and the SUSY particles have masses at or below the Fermi scale, i.e. < 1000 GeV but above 100 GeV. Indeed, the masses of the SUSY particles are expected to be of the same order as M_W, M_Z, and M_H.

The Big Bang Theory

The age of the universe is a function of the temperature T of the cosmic radiation radiated during the beginning of time or during the birth of the universe and is given by

$kT = (45\, \hbar^3 c^5\, /\, 32\, \pi^3\, G)^{1/4} \times 1/\, t^{1/2}$ or $kT = 1$ MeV $/\, t^{1/2}$

This equation for the age of the universe holds good for the early universe.

$k = 1.38 \times 10^{-23}$ joule per kelvin is Boltzmann constant.

$\hbar = h/2\pi = 6.582 \times 10^{-22}$ MeV-s is Reduced Plank constant.

$G = 6.67 \times 10^{-11}$ m^3 kg^{-1} s^{-2} is gravitational constant.

kT is the average energy per particle and t is the age of the universe in seconds.

at $t = 10^{-43}$ second, $kT = 3.16 \times 10^{21}$ MeV or 3.16×10^{18} GeV, $T = 3.7 * 10^{31}$ K

at $t = 10^{-39}$ second, $kT = 3.16 \times 10^{19}$ MeV or 3.16×10^{16} GeV, $T = 3.7 * 10^{29}$ K

at $t = 10^{-11}$ second, $kT = 3.16 \times 10^5$ MeV or 3.16×100 GeV, $T = 3.7 * 10^{15}$ K

at $t = 10^{-10}$ second, $kT = 100$ GeV, $T = 1.2 * 10^{15}$ K

at $t = 10^{-9}$ second, $kT = 31.6$ GeV, $T = 3.7 * 10^{14}$ K

at $t = 10^{-6}$ second, $kT = 1000$ MeV or 1 GeV, $T = 1.2 * 10^{13}$ K

at $t = 10^{-4}$ second, $kT = 100$ MeV, $T = 1.2 * 10^{12}$ K

at $t = 1$ second, $kT = 1$ MeV, $T = 1.2 * 10^{10}$ K

About 13.7 billion years ago, our universe was an infinitesimally small region of space of infinite density. This is called singularity. Due to some unknown reason, that point universe exploded (which

is called the Big Bang) and began to expand and that was also the beginning of time.

As the universe expanded from singularity and cooled, it underwent a series of transitions that are described by different era i.e. **Inflation, Particles Production, Nuclear Synthesis, Photon Decoupling, Cosmic Microwave Background Radiation (CMBR), the formation of cosmic structures** – galaxies, clusters of galaxies, and large-scale cosmic filaments.

Age of universe is 13.7 billion years (13 billion and 700 million years)
The observable universe is roughly 93 billion light-years in diameter.
The observable universe is the portion of the universe that we can see, as light from more distant regions hasn't had time to reach us.
The universe has been expanding since the Big Bang, stretching space and increasing distances between galaxies. Due to this expansion, the observable universe is larger than the age of the universe would suggest.

There are regions of space that are expanding away from us faster than the speed of light, making them unreachable. The true size of the universe beyond the observable part is unknown and could be infinite.

An object can not move faster than the velocity of light anywhere in the universe but the space between two points can expand faster than the velocity of light.

One second after the Big Bang, the observable universe was approximately 20 light-years in diameter whereas light can travel only 1 light second in one second.

Imagine a balloon with dots drawn on its surface. As you inflate the balloon, the dots move further apart, not because the dots are moving across the balloon's surface, but because the rubber of the balloon (space) is stretching between them.

Similarly, the space between distant galaxies is stretching, causing them to move away from us at an accelerating rate.

Due to the expansion of the universe, the distant objects are further away than their light travel time.

Suppose a star is 10000 light years away from the Earth today. Then the light emitted by the star today would reach the Earth after 10000 years. That is the light travel time would be 10000 years but by the time that light reaches to the Earth, due to the expansion of the universe, star would be say, 10200 light years away from the Earth which is more than its light travel time.

Proper distance:

The actual, physically measured distance between an object and an observer, taking into account the expansion of the universe.

Light-travel distance:

The distance the light has traveled from the object to the observer, essentially how far away the object was when it emitted the light.

Comoving distance:

Comoving distance factors out the expansion of the universe, giving a distance that does not change with time except due to local factors, such as the motion of a galaxy within a cluster.

Spectral lines from the distant galaxies are found to be red shifted which implies that the galaxies are receding from us which further implies universe is expanding.

The amount of redshift indicates the galaxy's recessional velocity i.e. how fast it's moving away from us.

For a particular galaxy, **the velocity of recession v** is proportional to its proper distance D from the Earth and is given by Hubble's law: $v = H_0D$, where H_0 is the Hubble constant and its topical value is about 70 km per sec per Mpc. 1 Mega parsec (1Mpc) = 3.09×10^{19} kms.

By measuring the velocity, astronomers can calculate the proper distance to the galaxy by using Hubble's law.

Age of the universe is given by $t = 1/H_0$. $H_0 = 70$ kms/s/Mpc gives the age of universe of about 13.4 billion years.

However, the Hubble constant isn't actually constant.

The age of universe is about 13.7 billion years.

Assuming matter to have been conserved, the matter density of the universe is proportional to $1/R^3$, where R is the expansion parameter and assuming radiation to be in thermal equilibrium, the energy density of radiation ρ_s varies as $1/R^4$ (although photon density like matter density varies as $1/R^3$). The extra factor of $1/R$ in energy density is due to the red shift.

This all implies **the energy density of radiation decreases faster than the matter density with the expansion of the universe** that is, **at early enough times,** when the size of the universe had been small, **the radiation must have been dominant** instead of matter, whereas today the matter density dominates.

As per quantum mechanics, minimum time interval that can be defined is **10^{-43} seconds** which is called **Planck Time.** Corresponding distance called **Planck Length** (velocity of light multiplied by Planck Time) is **10^{-35} meter** and distance less than this cannot be defined. These time intervals and distances may be thought of as quantum mechanical fluctuations, appearing out of nothing and disappearing of their own volition.

Thus, the earliest moment after the Big Bang or after the beginning of time that can be described is $t = 10^{-43}$ seconds.

The **planck epoch** is the earliest period of time in the history of the universe from zero to approximately 10^{-43} seconds (Planck Time).

During the **planck epoch, all the four forces**, i.e. gravity, strong force, electromagnetic force and weak force were **unified** that is, the coupling constants of all these forces had the same universal value. Nothing is known of this period.

At $t = 10^{-43}$ seconds, the planck epoch ended and the grand unification epoch began.

At the end of the Planck epoch i.e. at $t = 10^{-43}$ seconds, when kT was about 3×10^{18} GeV, <u>gravity separated</u> from the strong, electromagnetic and weak forces.

The **grand unification epoch** is the period of time in the history of the universe from $t = 10^{-43}$ to $t = 10^{-39}$ seconds.

During the **grand unification epoch**, the strong, electromagnetic and weak forces were **unified** that is, the coupling constants of all these three forces had the same value $\alpha_{GUT} = 1/24$.

At $t = 10^{-39}$ seconds, when kT was about 3×10^{16} GeV, <u>strong force separated</u> that is, the strong force became stronger than the weak and electromagnetic forces.

During $t = 10^{-39}$ to $t = 10^{-10}$ seconds, the electromagnetic and weak forces were **unified.**

This phase transition triggered the cosmic inflation.

During **inflationary epoch (t = 10^{-36}s and t = 10^{-32}s),** hot microscopic universe expanded exponentially which is called inflation. During inflation universe became super cooled.

After the end of inflation, the supercooled universe is reheated and reverts to the conventional hot Big Bang model.

During inflation, the volume of universe increased by a factor of at least 10^{78} (i.e. an expansion of distance by a factor of at least 10^{26} in each of the three dimensions)

After inflation, cosmic expansion **decelerated** to much slower rates, until around 9.8 billion years after the Big Bang (4 billion years ago) it began to gradually expand more quickly, and is still doing so.

Inflation smoothed out the universe and amplified quantum fluctuations into the large-scale structures we observe today.

After inflation, the universe was incredibly hot and dense, filled with fundamental particles and their antiparticles.

The small excess of quarks over antiquarks led to a small excess of baryons over antibaryons.

For every 1 billion antibaryons formed, 1 billion + 1 baryons formed.

At t = 10^{-10} seconds, when kT was about 100 GeV, <u>electromagnetic force separated </u> that is, the electromagnetic force became stronger than the weak force. (See the topic: Electroweak Unification)

As the temperature continued to decrease, the strong force got stronger and stronger and the weak force got weaker and weaker. The electromagnetic force also got weaker but at much less rate than the weak force.

During **quark epoch (t = 10^{-12}s to t = 10^{-6}s),** the universe was a quark-gluon plasma (QGP) where quarks and gluons moved freely because collisions were too energetic for them to bind.

During **hadron epoch (after ~10^{-6}s),** as the universe cooled further (below the binding energy of hadrons), quarks combined to form hadrons e.g. protons and neutrons, marking the end of the Quark Epoch.

In the primordial tiny universe, the average energy per particle was extremely large and all the particles were in thermal equilibrium, and thus existed in comparable numbers. For example, protons and photons were in thermal equilibrium through the reversible reaction p + p` $\rightleftharpoons$ γ + γ that is, when a proton converted into a photon, its total energy that is, the sum of the rest energy and kinetic energy of the proton converted into the total energy of a single photon. Now, when the photon began to convert into a proton, then during that little time interval, **due to the expansion of the universe, the universe became a little cooler and the average energy per particle decreased a little.** Thus, the photon converted into a proton of energy little less than the energy of the previous proton. Now, when the proton began to convert into a photon, then due to the further expansion of the universe, universe again became a little cooler and the average energy per particle again decreased a little and the proton transformed into a photon of even lesser energy and so on. Now, when the conversion of a proton into a photon was taking place at kT = M_pc^2 where M_p is the rest mass of the proton, then the proton converted into a photon of energy equal to the rest energy of the proton. Now, when the photon began to convert into a proton, then due to the further expansion of the universe and therefore further decrease in kT, the photon energy decreased below the rest energy of the proton, hence photon could not convert into a proton that is, the proton annihilated to the radiation. The same fate applied to all massive particles that is, **when the value of kT decreased below the rest energy of a particular particle, that particle disappeared.**

note: This is a simple explanation. If we consider statistical mechanics, then even at kT < rest energy of a particular particle, the small amount of those particles still survived in that universe.

At t = 10^{-10} second, kT = 100 GeV implies that the top quarks (rest mass: 173.3 GeV > 100 GeV) disappeared.

During t = 10^{-10} to t = 10^{-9} second, when kT decreased from 100 GeV to about 30 GeV, massive bosons, i.e. **Z^0 bosons** (rest mass: 91 GeV > 30 GeV) and **W^+, W^- bosons** (rest mass: 80 GeV > 30 GeV) **also disappeared.**

At t = 10^{-6} second, kT = 1000 MeV or 1 GeV implies that all the particles whose rest mass were larger than 1000 MeV or 1 GeV also disappeared after **t = 10^{-6} second.**

Thus, the **charm quarks** (rest mass: 1300 MeV > 1000 MeV) and the **bottom quarks** (rest mass: 4200 MeV > 1000 MeV) also disappeared. **All the baryons having charm and/or bottom quarks**, e.g. Λ_c^+ (udc), Λ_b^0 (udb), Σ_c^{++} (uuc), Σ_b^+ (uub), Ξ_c^0 (dsc), Ω_b^- (ssb), Ω_{ccc}^{++} (ccc) also disappeared. **J/ψ mesons** (rest mass = 3.1 GeV > 1 GeV) and **upsilon mesons** (rest mass: 9.46 GeV > 1 GeV) also disappeared. **D Mesons** (rest mass ~ 2 GeV) and **B Mesons** (rest mass > 5 GeV) also disappeared. All **sigma baryons** (rest mass ~ 1193 MeV > 1000 MeV), **lambda baryons** (1116 MeV), **Xi baryons** (1318 MeV), **delta baryons** (1232 MeV), **sigma-star baryons** (1385 MeV), **Xi-star baryons** (1530

MeV), **omega baryons** (1672 MeV), **phi mesons** (1020 MeV) **also disappeared.**

At t = 10^{-4} second, kT = 100 MeV implies that all the particles whose rest mass were larger than 100 MeV also disappeared after **t = 10^{-4} second.** Thus, the **strange quarks** (rest mass: ~ 100 MeV) and **muons** (rest mass: 105 MeV > 100 MeV) also disappeared. All the **kaons** (rest mass ~ 493 MeV > 100 MeV), **charged pions** (139 MeV), **neutral pions** (134 MeV), **eta mesons** (547 MeV), **eta prime mesons** (957 MeV), **K-star mesons** (892 MeV), **rho mesons** (770 MeV), **omega mesons** (782 MeV) **also disappeared.** Almost all the protons and neutrons also disappeared.

At t = 1 second, kT = 1 MeV implies that only electrons, electron type neutrinos v_e and muon type neutrinos v_μ survived but protons and neutrons are lightest baryons and stable enough to leave a small residue (one billionth residue). Thus, **after the first second of time, the end products of the Big Bang apart from predominant leptons** (electrons, electron-type neutrinos, muon-type neutrinos) **and photons were protons and neutrons.**

This explanation of the first second of the universe is based on the Big Bang theory which is strongly supported by the observation of Cosmic Microwave Background Radiation (CMBR). The spectral distribution of this radiation as measured with the COBE (Cosmic Microwave Background Explorer) satellite is exactly as that predicted for blackbody at T = 2.73 Kelvin. All the massive particles disappeared within the first second after the Big Bang. These highly unstable particles can now be produced in laboratories. They are also found in cosmic rays.

For every one billion + 1 protons produced, one billion antiprotons had formed just after big bang.

Thus, one billion protons and one billion antiprotons annihilated to 2 billion photons and 1 proton had no antiproton to be annihilated. That is out of total protons and antiprotons formed during the Big Bang, on billionth protons survived and all the protons in the universe today are due to that residue.

A similar process happened for electrons and positrons. That is annihilation of the electrons and positrons into photons left just one in 10^8 of the original electrons and none of the positrons.

Big Bang Nucleosynthesis (BBN) was a brief period from about 3 to 20 minutes after the Big Bang when the universe cooled enough for protons and neutrons to create primordial abundance of the light elements like hydrogen, helium. BBN produced about 75% hydrogen and 25% helium by mass.

After t ~ 1 second, neutrons combined with electron-type neutrinos to **produce protons** and electrons ($n + v_e \rightleftharpoons p + e^-$) and **protons** combined with electron-type antineutrinos to **produce neutrons** and positrons ($p + v_e` \rightleftharpoons n + e^+$) through the weak interactions. The **neutrons disappeared by beta-decay** ($n \rightarrow p + e^- + v_e`$) which is also a weak process.

If, only weak interactions and beta-decay had existed, then all the neutrons would disappear and the early universe would consist mainly of protons and electrons.

However, **after t ~ 1 second, neutrons and protons also started to form deuterons** through the electromagnetic interaction ($n + p \rightleftharpoons {}^2H + \gamma$) which is **faster** than the beta decay, hence **nucleosynthesis can begin before neutrons disappeared through beta decay.**

A deuteron is a nucleus of deuterium, which consists of one proton and one neutron.

Matter consisting of neutrons, protons and deuterons was in thermal equilibrium with the photons through reversible reaction n + p ⇌ ^{2}H + γ.

The binding energy of a deuteron is 2.22 MeV that is, when a neutron and a proton combine to form a deuteron, 2.22 MeV energy is released. A deuteron and a photon can convert into a neutron and a proton provided the photon has energy > 2.22 MeV.

The mean photon energy at temperature T is approximately 2.7kT. Thus, the photon energy would be equal to the binding energy of a deuteron i.e. 2.22 MeV, when kT ~ 0.82 MeV (2.7 kT = 2.7 × 0.82 MeV ~ 2.22 MeV).

Within a first second after the Big Bang, most of the electrons and protons disappeared and left only a small residue.

Thus, after the first second, the number density of photons exceeded one billion times the number density of matter particles. The distribution of the photons in that fireball was of blackbody type. This means, **even at KT (average energy per particle) much less than 0.82 MeV, there would exist some photons, each with energy > 2.22 MeV and only at a much lower temperature such that kT = 0.05 MeV (corresponding to t ~ 400 sec and T ~ 5.8 * 10^8 K), almost all the photons had energy < 2.22 MeV.**

Thus, after **t ~ 400s**, deuterons could not convert into neutrons and protons which means **deuterons became stable when kT was 0.05 MeV.**

****

kT = 1 MeV / t$^{1/2}$

0.05 MeV = 1 MeV / $t^{1/2}$ (if kT = 0.05 MeV)

$t^{1/2}$ = 1 MeV / 0.05 MeV = 20

t = 400 seconds

kT = 0.05 MeV

T = (0.05 * 10^6 eV) / k

T = (0.05 * 10^6 eV) / (1.38 × 10^{-23} joule per kelvin)

T = (0.05 * 10^6 * 1.6 * 10^{-19} joule) / (1.38 × 10^{-23} joule per kelvin)

T = (8 * 10^{-15} joule) / (1.38 × 10^{-23} joule per kelvin)

T ~ 5.8 * 10^8 K

****

Calculations show, **at kT = 0.87 MeV (t ~ 1.28 seconds), neutrons and protons went out of equilibrium** that is, those weak interactions ($n + v_e \rightleftharpoons e^- + p$ and $p + v_e` \rightleftharpoons e^+ + n$) took place no more and by that time, neutron-to-proton ratio was 0.23 that is, for every 100 protons, there existed 23 neutrons. Then, neutrons converted into protons through the beta decay ($n \rightarrow p + e^- + v_e`$). Thus, with time, the number of protons increased even more and the number of neutrons decreased even more and **at kT = 0.05 MeV = 5000 eV (t ~ 400 seconds, T ~ 5.8 * 10^8 K), neutron-to-proton ratio was 0.14** that is, for every 100 protons, there were 14 neutrons.

At kT = 0.05 MeV or t ~ 400 sec, deuterium (consisting of one proton and one neutron) had become stable,

Once **deuterium** (consisting of one proton and one neutron) became stable, it **combined with a neutron and** then with **a proton** or with a proton and then with a neutron **to form helium atom** (having two protons and two neutrons).

Thus, **after t ~ 400 sec, neutrons were bound inside the deuterons or heavier nuclei** and therefore no longer decayed and **neutron-to-proton ratio (r = 0.14) became fixed.**

After t ~ 1 second, neutrons and protons started to form deuterons. Similarly, **after t = 1 second, electrons and protons started to produce hydrogen atoms.**
Matter consisting of electrons, protons and hydrogen atoms was in thermal equilibrium with the photons through reversible reaction $e^- + p \rightleftharpoons H + \gamma$.
The binding energy of a hydrogen atom in its ground state is 13.6 eV that is, when an electron and a proton combine to form a hydrogen atom, 13.6 eV energy is released. A hydrogen atom and a photon can convert into an electron and a proton provided the photon has energy > 13.6 eV (so that the electron of the hydrogen atom by absorbing that energy of photon become free from the attraction of the proton).
The mean photon energy at temperature T is approximately 2.7kT. Thus, the photon energy would be equal to the binding energy of a hydrogen atom i.e. 13.6 eV, when kT ~ 5 eV (2.7 kT = 2.7 × 5 eV ~ 13.6 eV).

After the first second, the number density of photons exceeded one billion times the number density of matter particles. The distribution of the photons in that fireball was of blackbody type. This means, **even at KT (average energy per particle) much less than 5 eV, there would exist some photons, each with energy > 13.6 eV and only at a much lower temperature such that kT = 0.3 eV (corresponding to t ~ 1.2 × 10¹³ sec or 3.8 × 10⁵ years i.e. 380000, T ~ 3400 K) almost all the photons had energy < 13.6 eV.** Thus, after **t ~ 3.8 × 10⁵ years,** hydrogen atom could not convert into electron and proton which means hydrogen atom (H) and radiation (γ) decoupled and hydrogen atom became stable **when kT was 0.3 eV.**

This also implies that **the matter density and the radiation density became equal at about 3.8×10^5 years i.e. when kT was 0.3 eV.** Thereafter, matter density started to dominate the energy density of the universe.

****

kT = 1 MeV / $t^{1/2}$

0.3 eV = 1 MeV / $t^{1/2}$ (if kT = 0.3 eV)

$t^{1/2}$ = 1 MeV / 0.3 eV = 10^7 / 3

t ~ 1.2 * 10^{13} seconds

t ~ 1.2 * 10^{13} / 31557600 years

t ~ 3.8 * 10^5 years

kT = 0.3 eV

T = (0.3 eV) / k

T = (0.3 eV) / (1.38 × 10^{-23} joule per kelvin)

T = (0.3 * 1.6 * 10^{-19} joule) / (1.38 × 10^{-23} joule per kelvin)

T = (3 * 1.6 * 10^{-20} joule) / (1.38 × 10^{-23} joule per kelvin)

T = (4.8 * 1000 joule) / (1.38 joule per kelvin)

T = (4800 joule) / (1.38 joule per kelvin)

T ~ 3400 K

****

A deuteron is the nucleus of a deuterium atom. Deuteron is composed of one proton and one neutron.

Deuterium means deuteron and an electron.

Deuterium (D_2O) is an isotope of hydrogen and has an atomic mass roughly twice that of hydrogen.

Deuterium nuclei i.e. deuterons fused with other protons and neutrons to create helium nuclei. Helium nucleus consists of 2 protons and 2 neutrons.

These newly formed helium nuclei then captured electrons from the surrounding plasma to create helium atoms. Helium atom consists of helium nucleus and two electrons.

Big Bang nucleosynthesis mainly formed hydrogen and helium atoms.

After Big Bang nucleosynthesis, hydrogen and helium gas clouds acted as the initial matter for the star formation and nuclear fusion in the stars formed heavier elements. This is called stellar nucleosynthesis.

The Big Bang Nucleosynthesis theory predicts that roughly 25% mass of the universe consists of helium. The helium mass fraction, i.e. the mass abundance of helium relative to hydrogen is calculated to be equal to $2r/(1 +r) = 0.25$.

The helium mass fraction measured in solar system, in globular cluster, etc. is approximately 0.24. **Nearly the same value of the calculated and observed helium mass fraction supports the Big Bang theory.**

Cosmic Microwave Background Era, Population III Stars

Cosmic Microwave Background Era

Up to 380000 years after the Big Bang (at a redshift of z = 1100), universe was a hot opaque plasma because due to the frequent collisions of photons with electrons and protons, photons could not travel freely.

3.8 × 10⁵ years after the Big Bang when the universe had cooled to ~3400 Kelvin and average energy per particle kT was 0.3 eV, the reversible reaction $e^- + p \rightleftharpoons H + \gamma$ no longer took place that is hydrogen atom could not convert into electron and proton which means hydrogen atom (H) and radiation (γ) decoupled and hydrogen atom became stable.

This era when electrons and protons combined to form neutral stable hydrogen atoms is called decoupling or recombination.

Decoupling or recombination allowed light to travel freely, ending its constant scattering by free electrons and protons.

Photon decoupling happened 3.8×10^5 years after the Big Bang and hence 380000 years after the Big Bang universe became transparent to light, and the cosmic microwave background radiation, the afterglow of the hot Big Bang, was emitted. to light escaping a star, marking the farthest back we can "see" with light.

Recombination happened everywhere in the universe. Thus, we see CMB radiation coming from all directions.

The CMB is faint microwave radiation, a relic of the hot, dense early universe, detectable from every direction in the sky.

The CMB era marks the universe's transition from an opaque plasma to a transparent, cooler state, allowing the photons from that moment to travel freely.

The **Cosmic Dark Ages** was a period 380,000 years (z ~1100) after the Big Bang when the universe was filled with hydrogen and helium gas but lacked luminous stars and up to 100 million years (z ~30) when the first stars - free of any metals (Population III) - formed, emitting UV light that reionized the surrounding gas.

Ending of the Dark Ages was the beginning the Epoch of Reionization. Epoch of Reionization was a period during 100 million years after the Bing Bang and up to 1 billion years after the Big Bang.

Population III Stars

Population III.1 (Pop III.1) stars are the first generation of stars i.e. the very first stars or the initial, purely primordial stars in the universe often forming in clusters. They had formed from pristine gas (mostly hydrogen and helium) after the Big Bang. Thus, they were metal-free, extremely massive, very hot, luminous (emitting UV radiation), and had very short lifespans, only a few million years. Lives of typical Pop III.1 stars ended in the form of supernovae explosions (supernovae) or gamma-ray bursts (GRBs).

Heavier elements act as coolants and decrease the mass of the stars.

Lack of heavier elements in Pop III.1 means lack of coolants in Pop III.1. Thus, they were very massive. Masses of Pop III.1 stars may be up to 1000 solar masses given that the primordial gas does not contain any efficient coolants in star-forming regions.

Supernovae explosions of Pop III.1 stars seeded the universe with the first heavy elements (metals), enabling the formation of later, less massive stars (like our Sun) and planets.

Pop III.2 are also primordial stars i.e. formed from pristine gas (mostly hydrogen and helium) after the Big Bang but they contained some coolants i.e. heavy elements in their star-forming regions because supernovae explosions at the end of Pop III.1 stars had enriched the star-forming regions with some heavy elements. Pop III.2 had lower masses (around 40-60 solar masses) than Pop III.1 because of coolants in their star-forming region.

Population II are old, metal-poor stars, found in galactic halos and globular clusters.

Population I are young, metal-rich stars (e.g., Sun).

Formation of Super Massive Black Hole (SMBH)

Black hole seeds from the remnants of typical Pop III star supernovae don't grow fast enough and hence cannot become SMBH.

However, supermassive Pop III stars (more than 260 solar masses) can collapse directly into black holes of similar mass and can grow into SMBH.

Supermassive Pop III stars burn helium in their cores, producing carbon. The carbon leaks into a surrounding shell where hydrogen is burning.The carbon combines with hydrogen to create nitrogen through the carbon/nitrogen/oxygen (CNO) cycle. Convection currents distribute the nitrogen throughout the star. Eventually, this nitrogen-rich material is ejected into space. The process continues for millions of years during the star's helium-burning phase, creating the nitrogen excess in the galaxy having such a star.

When such a supermassive star dies, it doesn't explode. Instead, it collapses directly into massive black hole seeds (100 to 10000 solar masses).

Observation of the nitrogen excess i.e. high N/O ratio (0.46) in the galaxy GS 3073 implies that the SMBH at the center of this galaxy might have formed from Pop III supermassive star existing in the early universe. This also implies SMBHs at the center of the quasars found in early universe also might have formed from Pop III supermassive stars. These stars lived briefly (around 250,000 years) but enriched the universe with heavy elements, laying foundations for later galaxies and seeding supermassive black holes.

The high N/O ratio in early galaxies is a key signature of massive Population III stars forming and dying via direct collapse, acting as seeds for the first black holes and enriching the universe with nitrogen.

Direct Collapse Black Hole (DCBH)

Supermassive black holes have already been observed at redshift z~7 and even more.

Direct Collapse Star (DCS) / Direct Collapse Black Hole (DCBH) is a specific, rapid process of direct collapse to form supermassive black hole (SMBH) seeds (10000 to 1 million solar masses) at high redshifts i.e. in the early universe.

Key conditions for the direct collapse are low metallicity and the **suppression of H_2 cooling by UV radiation** from the nearby galaxies. **Hot gas with few heavy elements (metals) cools less efficiently.**

H_2 molecules absorb thermal energy from gas and re-emit it as infrared photons, which escape the gas cloud, causing cooling.

However, the strong UV radiation (Lyman-Werner bands) from nearby forming stars or galaxies destroys molecular hydrogen (H_2), a crucial coolant, of the gas cloud. Thus, the gas remains hot.

The massive hot gas cloud doesn't break into smaller pieces and hence instead of becoming typical Pop III stars or supermassive Pop III star first, the entire gas cloud collapses into massive black hole seed, potentially 10,000 to 1 million solar masses.

A supermassive Pop III star collapses into black hole seeds of 100 to 10000 solar masses, whereas direct collapse black hole process produces a supermassive black hole seed of 10000 to 1 million solar masses,

These SMBH seeds of 10000 to 1 million solar masses putatively formed within the redshift range z = 15–30, when the Universe was about 100–250 million years old.

These seed black holes then grow rapidly by accreting matter from the surrounding and become SMBHs in relatively much less time.

About the Author

Mohit Joshi, The Author Of This Book Is Graduate In 'Electronics & Communication Engineering' & Is GATE (Graduate Aptitude Test In Engineering) Qualified. He Is Particularly Interested In Cosmology & High Energy Physics. Albert Einstein & Richard Feynman Are His Favourite Nobel Laureates.

Connect with Mohit

Amazon Author Page - https://www.amazon.com/author/mohitjoshi

****

www.ingramcontent.com/pod-product-compliance
Lightning Source LLC
Chambersburg PA
CBHW051852170526
45168CB00001B/75